LIFE IS A SIMULATION- MASTER THE CODE, CONTROL YOUR REALITY

BY
Pauline Stoltzner
PhD., MSN, APRN, FNP-BC, PMHNP-BC, CSAP

TABLE OF CONTENTS

Introduction: Welcome to the Simulation

On a quiet Tuesday in 2003, Oxford philosopher Nick Bostrom published a paper that would shake the foundations of reality itself. His simulation hypothesis argued that at least one of three propositions must be true: humanity will go extinct before reaching technological maturity, advanced civilizations choose not to run simulations of their ancestors, or we are almost certainly living in a computer simulation.

The logic was elegantly simple yet profoundly unsettling. If civilizations survive long enough to develop advanced computing power, and if they choose to run simulations of their ancestors, much like we do with video games, but infinitely more sophisticated, then the number of simulated beings would vastly outnumber "real" beings. If there are thousands of simulated versions of you for every original version, then statistically, you're more likely to be one of the simulations.

Elon Musk took this reasoning even further during a 2016 conference, arguing that the odds we're living in base reality are "billions to one." His reasoning was deceptively straightforward: "Forty years ago, we had Pong. Two rectangles and a dot. That was what games were. Now, forty years later, we have photorealistic, 3D simulations with millions of people playing simultaneously, and it's getting better every year. Soon we'll have virtual reality, augmented reality. If you assume any rate of improvement at all, then the games will become indistinguishable from reality."

Tech leaders across Silicon Valley and scientists at MIT have joined the conversation. Neil deGrasse Tyson puts the odds at fifty-fifty. Sam Altman,

CEO of OpenAI, has stated there's a "pretty good chance we're in a simulation." Rizwan Virk at MIT has written extensively about what he calls the more-likely-than-not possibility that our entire universe is nothing more than an elaborate program running on some advanced civilization's hardware.

But here's what none of them are talking about, and it's the reason I've written this book: If we're living in a simulation, then reality has code, and if it has code, it can be hacked.

My name is Dr. Pauline Stoltzner, and I've spent the last two decades discovering and exploiting the glitches in our reality simulation. I've used these techniques to hack my way from sleeping in tents and scrounging for food to running a successful medical practice, earning multiple degrees including a PhD, and creating the life I once thought was mathematically impossible for someone from my background.

This isn't another book about positive thinking or the law of attraction dressed up in new language. This is about finding actual exploits in the code that runs your personal reality and learning to manipulate them for measurably better outcomes. I'm not talking about wishful thinking, I'm talking about systematic techniques grounded in neuroscience, psychology, and systems theory that allow you to reprogram the fundamental operating system generating your life experience.

The evidence that something like this is possible surrounds you every day. Why do some people seem to have "cheat codes" for life, consistently attracting opportunities and success, while others work just as hard but barely keep their heads above water? Why do certain individuals recover from devastating setbacks within months while others never recover at all? Why do some people radiate confidence and capability while others, equally talented, shrink from their own potential?

The conventional answer is luck, genetics, or privilege. But I've discovered something different. These people have learned, consciously or unconsciously, to hack the underlying code of their personal reality simulation. They're running different programming, and it generates completely different outcomes.

The Personal Reality Simulation

Whether we're living in a literal universal computer simulation remains an open philosophical question. But here's what's not debatable: your personal experience operates exactly like programmable code and understanding this changes everything.

Consider what your brain actually does. Right now, it's processing roughly eleven million bits of information per second from your sensory organs, sight, sound, touch, smell, taste, proprioception. But here's the startling part: only about forty bits of that information ever reaches your conscious awareness. That means your brain is filtering out 99.9996% of available reality before you ever experience it.

This isn't random filtering. Your brain acts like a sophisticated software program, deciding what information to show you based on your beliefs, expectations, values, and past experiences. It literally constructs your reality by choosing what data makes it past the filter and what gets discarded as irrelevant.

Change the filtering criteria, your beliefs about what's possible, what you deserve, what's appropriate for "people like you," and you change the reality you experience. Not metaphorically. Actually.

Dr. Joe Dispenza's research at the University of California demonstrates that this filtering system operates exactly like computer code. When your belief system says, "opportunities are scarce," your brain filters out most opportunities and shows you primarily problems and obstacles. When your programming says, "I'm not smart enough for success," your brain interprets challenges as confirmation of inadequacy and automatically avoids situations requiring high performance.

This happens below conscious awareness, which is why most people never realize their limitations aren't real, they're just programming executing automatically.

The breakthrough that changed my life came at age six, hiding in a tent while my parents fought about money they didn't have, for the thousandth time. In that moment of desperation, something shifted in my consciousness. Instead of feeling like a victim trapped in chaos, I suddenly saw their behavior differently, not as random cruelty or bad luck, but as predictable code execution.

My mother's addiction wasn't a character flaw; it was programming installed during her traumatic childhood. My father's rage wasn't about me; it was an automatic response pattern triggered by feelings of powerlessness. Their poverty wasn't bad luck; it was the inevitable output of scarcity programming running on infinite loop.

More importantly, I realized I wasn't part of their program. I was running my own system, and I could choose what code to execute.

This was my first glitch experience, a moment when normal reality suspended and I could see the underlying structure of how consciousness creates experience. Most people have these moments and dismiss them as strange thoughts or temporary clarity. I recognized mine as something more: a window into the programmable nature of reality itself.

From that night forward, I began treating my life differently. Instead of accepting limitations as permanent features of reality, I looked for the code generating those limitations. Instead of feeling helpless about circumstances, I searched for ways to modify the programming creating those circumstances. Instead of being a victim of inherited patterns, I became, slowly, clumsily, but deliberately, the programmer of my own experience.

This book contains everything I've learned about that process. Not just concepts and theories, but specific techniques you can implement immediately to start hacking your own reality simulation.

What Makes This Different

Walk into any bookstore and you'll find hundreds of self-help books promising transformation through positive thinking, visualization, or mindset shifts. Most of them fail because they're trying to change outputs without understanding or modifying the underlying code.

It's like trying to fix a software bug by clicking your mouse harder. The problem isn't your effort, it's that you're working at the wrong level of the system.

Real transformation happens at the programming level, where your core beliefs and automatic patterns generate your experience. Once you understand how your personal reality simulation actually works, the specific code creating your current results, you can systematically debug limiting programs and install upgraded ones that generate better outcomes.

This book will teach you to identify the exact code creating unwanted results in your life, debug limiting beliefs that function like software bugs, install new programming that generates better outcomes automatically, exploit glitch states where rapid transformation becomes possible, and ultimately become the master programmer of your own existence.

The techniques I'll share aren't just my personal discoveries. They're grounded in solid scientific research from neuroscience, psychology, quantum physics, and systems theory. I've spent years studying how the brain creates reality, how beliefs generate experiences, and how consciousness interfaces with the physical world. More importantly, I've tested everything on myself first and then with hundreds of clients, students, and research subjects.

What I've discovered is that reality is far more malleable than most people imagine, but only when you know how to work with its underlying code structure.

My First Reality Hack

The first time I deliberately hacked my reality came during my medical training, in a moment of such desperation that I had nothing left to lose. I was about to take a big exam for a science class, despite studying more hours than I'd ever studied anything in my life, the concepts felt out of my reach.

The problem wasn't effort. The problem was code. My programming said: "You're not good at science. You don't have the kind of brain that understands chemistry. People from chaotic backgrounds like yours don't go into medicine. You're going to fail and prove everyone right about your limitations."

This wasn't just negative thinking I could overcome with affirmations. These were deep programs installed through years of struggle, reinforced by every teacher who gave up on me, every test I failed while my home life fell apart, every person who looked at my background and saw limited potential.

The night before the exam, I pulled an all-night study session out of pure desperation. Around three in the morning, exhausted and overwhelmed, something extraordinary happened. My normal perception suddenly shifted.

Instead of seeing pathophysiology as a foreign language I'd never master, I saw it differently. The molecular structures stopped being incomprehensible abstractions and became beautiful, logical patterns. I could see relationships between different compounds as clearly as I could see relationships between people. The entire subject reorganized itself in my mind into something elegant and comprehensible.

More profoundly, my identity shifted. I wasn't "someone who can't understand the science," I was just a conscious being engaging with information. The limitation wasn't real. It was programming, and in that moment, the programming had temporarily suspended.

Time distorted. I studied intensely for six more hours, but it felt like thirty minutes. Effort became effortless. Complex concepts that had been impossible suddenly made perfect sense. I could see connections and patterns I'd never noticed before. My usual self-doubt and anxiety completely disappeared, replaced by calm confidence and curiosity.

I didn't just pass the final exam; I scored in the top five percent of the class. But the real breakthrough wasn't the grade. It was the proof that my limitations were optional. They were code that could be suspended and rewritten.

From that experience, I began actively studying what I now call "glitch states," moments when your normal programming temporarily goes offline and rapid transformation becomes possible. I learned to recognize them, trigger them deliberately, and most importantly, exploit them to install permanent upgrades to my reality operating system.

Every major breakthrough in my life since then has involved recognizing, accessing, or creating these glitch states where normal limitations don't apply. The question isn't whether you can hack your reality. The question is whether you're ready to learn how.

The Scientific Foundation

Before we dive into specific techniques, you need to understand the scientific foundation that makes reality hacking possible. This isn't mysticism or wishful thinking dressed up in simulation language. It's grounded in legitimate research across multiple fields.

Neuroscience research on neuroplasticity shows that your brain literally rewires itself throughout your entire life based on your thoughts, experiences, and where you focus your attention. Dr. Rick Hanson's work at UC Berkeley demonstrates that "neurons that fire together, wire together," meaning you can physically change your brain structure through mental practice alone. Studies of London taxi drivers show their hippocampus actually grows larger as they learn the city's street layout. Meditation research proves that just eight weeks of mindfulness practice creates measurable changes in brain structure.

This neuroplasticity is the hardware foundation that makes software updates possible. When you change your mental programming, your brain physically reorganizes to support those changes.

Psychology research on the placebo effect demonstrates that belief alone can create measurable physical changes. In clinical trials, patients who believe they've received powerful medication often experience the same benefits as those who actually received the drug. If belief can literally alter brain chemistry, immune function, and physical healing, it can certainly change life circumstances by affecting what you notice, how you interpret events, and what actions you take.

Cognitive Behavioral Therapy research shows that changing thought patterns creates predictable, repeatable changes in emotions and behavior. This isn't theory, it's documented, peer-reviewed science with decades of clinical validation.

Systems theory research demonstrates how small changes in complex systems can create large-scale transformations through feedback loops and emergence. Your consciousness operates as what scientists call a "complex adaptive system," meaning small changes to core programming can cascade into massive life transformations.

While quantum physics shouldn't be oversimplified or misused, legitimate research on the observer effect shows that the act of observation changes the behavior of subatomic particles. Whether or not consciousness literally creates physical reality at the quantum level, it certainly determines which aspects of reality you notice, respond to, and act upon, which for practical purposes amounts to the same thing.

Your Reality Hacking Journey

This book is structured as a systematic progression from recognizing that you're running inherited programming to becoming the conscious master programmer of your own reality simulation.

Part One teaches you to see the code currently running your reality. Most people are completely unconscious of their programming, experiencing its outputs as "just how life is" or "just who I am." You'll learn to recognize your beliefs, patterns, and automatic responses as programmable code rather than fixed characteristics. You'll discover how to identify glitch states, moments when rapid transformation becomes possible, and learn to exploit them for permanent upgrades.

Part Two provides specific techniques for hacking the core systems generating your reality: your money code, relationship algorithms, career programming, and health systems. For each area, you'll learn to identify limiting code, debug it systematically, and install upgraded programming that generates better results automatically. These aren't vague concepts, they're step-by-step protocols with specific daily practices, measurable milestones, and real-world case studies.

Part Three covers advanced techniques for emotional programming, master-level reality manipulation, and sophisticated consciousness hacking. You'll learn to create and maintain peak performance states, eliminate limiting emotional patterns, and consciously design your reality rather than just reacting to circumstances.

Part Four guides your transition from user to administrator of your own reality simulation. You'll develop the ability to make real-time adjustments to your programming, help others upgrade their code, and use your mastery for positive impact in the world. This is about becoming not just successful, but significant, using your reality hacking skills to contribute to something larger than yourself.

Every chapter includes practical exercises, implementation protocols, and troubleshooting guides for common problems. This is a technical manual for hacking human consciousness, not a collection of inspirational stories and vague advice.

What This Book Requires From You

I won't lie to you: this work requires commitment. You're not going to read this book once, feel inspired, and wake up tomorrow with a transformed life. Reality hacking is a skill that must be practiced consistently until new programming becomes automatic.

You'll need to spend ten to fifteen minutes daily on programming exercises. You'll need to take actions that feel uncomfortable because they're outside your current programming. You'll need to persist through the inevitable moments when old code tries to reassert itself and pull you back to familiar patterns.

You'll need to be willing to question everything you've believed about yourself and what's possible. You'll need courage to implement changes even when people around you resist or criticize. You'll need patience with the process, understanding that real programming changes take weeks and months of consistent practice, not days.

Most importantly, you'll need to take full responsibility for your results. No more blaming circumstances, genetics, or bad luck. No more waiting for external conditions to change before you can be successful or happy. You're the programmer now, which means you're responsible for the code you're running and the reality it generates.

If you're ready for that level of commitment and responsibility, this book will give you tools more powerful than anything you've encountered before. If you're looking for easy answers or quick fixes, you should probably stop reading now.

The Invitation

The simulation has been running your entire life, executing code installed by your family, culture, and experiences, code you never consciously chose. Most people live their entire lives without realizing they have admin access to their own system.

But you're reading this book, which means some part of you already knows the truth: your current limitations aren't fixed features of reality. They're programming, and programming can be hacked.

The techniques in this book work. I know because they transformed my life from poverty and chaos to success and fulfillment. I know because I've taught them to hundreds of people who've used them to create their own transformations. I know because they're grounded in solid science, not wishful thinking.

The simulation is waiting for your input. The code is ready to be rewritten. Your new reality is just a few programming sessions away.

The question isn't whether reality is hackable. The question is: are you ready to become the hacker?

Welcome to your reality hacking journey. Your real life is about to begin.

PART 1: DISCOVERING THE SIMULATION

Weeks 1-2: Learning to See the Code

Chapter 1: The Science of Living in Code

The simulation hypothesis isn't the fevered imagination of science fiction writers or tech billionaires with too much time on their hands. It's a serious academic theory with profound implications for how we understand reality, consciousness, and the nature of human experience, and whether or not we're literally living in a computer simulation. Understanding why brilliant scientists and philosophers take this idea seriously will fundamentally change how you think about your life and what's possible.

Nick Bostrom's original paper, published in the *Philosophical Quarterly* in 2003, didn't actually argue that we're definitely living in a simulation. His argument was more subtle: he demonstrated that at least one of three propositions must be true, humanity will go extinct before reaching technological maturity, posthuman civilizations choose not to run ancestor simulations, or we are almost certainly living in a simulation.

The brilliance of Bostrom's trilemma is that all three options seem unlikely, yet one must be true. Think about it: we haven't gone extinct yet and seem to be making steady technological progress, so option one seems pessimistic. Why would advanced civilizations with unlimited computing power choose not to run simulations of their ancestors, when we already run primitive simulations (video games, models, forecasts) with our limited technology? That makes option two seem implausible. Which leaves option three, the one that seems craziest but might actually be most likely.

The mathematics are compelling and disturbing. If there's any substantial chance that civilization reaches a posthuman stage and runs many ancestor simulations, then statistically, you're more likely to be in one of those simulations than in the original reality. If just one advanced civilization decides to run a thousand detailed simulations of human history, there would be a thousand simulated versions of you for every "real" version. The odds would be 1,000 to 1 that you're simulated.

Elon Musk extended this logic with his characteristically blunt assessment: given the rate of technological advancement, from Pong to photorealistic VR in just forty years, simulations indistinguishable from reality are inevitable if civilization survives. Therefore, the odds we're in the original, "base" reality rather than one of countless simulations are "billions to one."

But here's where it gets interesting for our purposes: whether or not we're in a literal computer simulation, your personal experience operates exactly like programmable software, and once you understand that, everything changes.

How Your Brain Programs Reality

Your brain doesn't show you reality as it actually is. It can't, there's far too much information. Instead, it constructs a user-friendly interface that shows you a simplified, interpreted version of reality, optimized for your survival and success.

Dr. Donald Hoffman's research at UC Irvine demonstrates this conclusively. His interface theory of perception shows that natural selection didn't evolve our senses to show us truth; it evolved them to show us what's useful for survival. Just as your computer desktop shows you files and folders rather than the underlying machine code, your consciousness shows you a constructed reality rather than raw sensory data.

Here's the astonishing part: your brain processes roughly eleven million bits of information per second from all your sensory organs. But cognitive science research shows that only about forty bits ever reach your conscious awareness. That means your brain is filtering out 99.9996% of available reality before you experience anything.

This filtering isn't random. It's governed by your beliefs, expectations, values, and past experiences, what I call your "reality code." Your brain literally constructs your experienced reality by deciding what information to show you based on your programming.

This is why two people can witness the exact same event and have completely different experiences of what happened. They're running different code, so their brains filter and interpret the same raw data differently. The event itself is neutral until it is processed through personal programming.

Think about how this plays out in real life. If your programming says, "the world is dangerous," your brain automatically scans for threats, filters out safety cues, and interprets ambiguous situations as threatening. You experience a dangerous world not because the world is objectively more dangerous for you than for others, but because your filtering system is programmed to show you danger.

If your code says, "I'm not smart enough for success," your brain interprets challenges as confirmation of inadequacy, filters out evidence of your capabilities, and automatically avoids situations requiring high performance. You experience being not-smart-enough, not because it's objectively true, but because your programming generates that experience.

This isn't just psychology, it's neuroscience. Brain imaging studies show that people with different belief systems exhibit distinct neural activation patterns when processing the same information. Your beliefs physically change how your brain processes reality.

The Neuroscience of Reprogramming

Until recently, scientists believed adult brains were essentially fixed, that after childhood, you were stuck with whatever neural wiring you'd developed. We now know this is completely false, and understanding why changes everything about what's possible for personal transformation.

Neuroplasticity research proves that your brain continuously rewires itself throughout your entire life based on your thoughts, experiences, and where you focus your attention. Dr. Rick Hanson's work at UC Berkeley shows that "neurons that fire together, wire together." When you repeatedly think certain thoughts or have certain experiences, you strengthen the neural pathways supporting those patterns.

This creates what Hanson calls "positive neuroplasticity," the ability to consciously rewire your brain for better outcomes. You're not stuck with the programming installed during childhood. You can deliberately create new neural pathways that support the reality you want to experience.

The evidence for this is overwhelming. Studies of London taxi drivers show that the hippocampus, responsible for spatial navigation, actually grows larger as they learn the city's complex street layout. Musicians develop enlarged auditory cortex regions. Meditation practitioners develop thicker prefrontal cortices and larger insulae. Even learning to juggle can lead to measurable changes in brain structure within weeks.

Your brain is constantly updating its hardware based on your software, your thoughts, beliefs, and mental habits. Run different mental programming, and your brain physically reorganizes to support it. This is the biological foundation that makes reality hacking possible.

Dr. Joe Dispenza's research demonstrates this even more dramatically. His work with advanced meditators shows that focused mental practice alone, without any physical changes to circumstances, can create measurable improvements in health markers, emotional states, and even gene expression. Your thoughts literally reprogram your biology.

But here's what most people miss: this reprogramming happens automatically, whether you're conscious of it or not. Your brain is always learning, always adapting, always reinforcing the patterns you run most frequently. The question isn't whether you're programming your brain; you are, constantly. The question is whether you're doing it consciously and deliberately, or unconsciously and randomly.

Most people are running inherited code on autopilot, reinforcing limiting beliefs and patterns without ever questioning them. Their brains are faithfully strengthening neural pathways for anxiety, self-doubt, and limitation because that's what they practice most consistently.

Reality hacking means taking conscious control of this process, deliberately installing and practicing new patterns until your brain rewires itself to support them automatically.

The Belief-Reality Feedback Loop

The relationship between beliefs and reality isn't one-directional. Your beliefs don't just filter your perception of reality; they actively create experiences that reinforce themselves through a powerful feedback loop.

Here's how it works: Your beliefs determine what you pay attention to and how you interpret it. Your interpretations determine how you feel and what actions you take. Your feelings and actions determine what results you get. Your results provide evidence that reinforces your original beliefs.

This feedback loop operates below conscious awareness, which is why it's so powerful and so hard to break without understanding its structure.

Example: Sarah believes "I'm not good at relationships." This belief programs her to:

Notice moments when connections feel awkward or difficult while filtering out positive interactions. Interpret ambiguous situations negatively; if someone doesn't text back immediately, she assumes they're losing interest rather than just busy. Feel anxious and insecure in relationships, which affects how she shows up. Behave in ways that create distance or trigger conflict, like excessive reassurance-seeking or preemptive rejection. Experience relationships that confirm her belief that she's not good at them.

The belief creates experiences that confirm the belief, which strengthens the belief, which creates more confirming experiences. It's a perfect self-reinforcing loop, a program running flawlessly to generate predictable outputs.

The psychological term for this is "confirmation bias," your brain's tendency to notice evidence supporting your beliefs while filtering out contradictory information. But it's more than just selective attention. Your beliefs actually cause you to behave in ways that create confirming experiences.

Psychologists call this a "self-fulfilling prophecy." If you believe you're unlikable, you unconsciously behave in ways that make people less likely to like you, avoiding eye contact, keeping interactions superficial, expecting rejection, and defending against intimacy. Then, when people don't connect with you deeply, it confirms your belief that you're unlikable.

This feedback loop explains why some people seem to have persistent "bad luck" across all areas of life. They're not actually unlucky; they're running code that generates unlucky experiences, which reinforces the code, which generates more unlucky experiences. They're trapped in a negative feedback loop their programming creates and maintains.

The good news is that positive feedback loops work the same way. People who "always land on their feet" aren't just lucky; they're running code that helps them notice opportunities, interpret challenges optimistically, maintain confidence during setbacks, take actions that create positive outcomes, and therefore experience results that reinforce their belief that they always land on their feet.

Reality hacking works by deliberately interrupting negative feedback loops and creating positive ones. When you change a core belief, it changes what you notice, how you interpret events, how you feel, what actions you take, what results you get, and therefore what evidence reinforces your beliefs.

Change the code, and the entire loop shifts to support your upgraded programming instead of your limiting programming.

The Placebo Effect as Evidence

One of the most compelling pieces of evidence that consciousness programs physical reality is the placebo effect, the well-documented phenomenon where belief alone creates measurable physical changes.

In clinical trials, patients receiving inert sugar pills often experience the same benefits as those receiving actual medication, not just subjective improvements in how they feel, but objective, measurable changes in body chemistry, immune function, and healing rates.

This effect is so powerful and so consistent that modern medicine must design elaborate double-blind studies to account for it. Pharmaceutical companies spend billions trying to prove their drugs work better than a placebo, and many drugs fail to beat the placebo effect.

Think about what this means: your beliefs about what you're receiving can literally change your brain chemistry, hormone levels, immune response, and physical healing. If belief can do that, it can certainly influence the opportunities you notice, the actions you take, and the results you create in your life.

The placebo effect isn't limited to medicine. Research shows similar phenomena across many domains:

Students who believe they're drinking alcohol (but actually receive non-alcoholic drinks) demonstrate impaired performance matching actual alcohol intoxication. Hotel housekeepers who learn their work counts as exercise show measurable improvements in weight, blood pressure, and body fat percentage without any change in their actual activities, just their belief about what they're doing. Athletes who believe they're using performance-enhancing substances (but receive placebos) show improved performance. People who believe they're receiving fake electric shocks show physiological stress responses identical to receiving real shocks.

Your beliefs don't just affect your subjective experience; they create objective, measurable changes in your body and behavior. This is the biological mechanism underlying reality hacking.

Quantum Physics and Consciousness (Used Carefully)

I want to address quantum physics carefully because it's often misused in self-help literature to make claims that aren't scientifically supported. I won't tell you that consciousness creates physical reality at the fundamental level, or that your thoughts directly manifest things from quantum possibilities, or that quantum mechanics proves mystical beliefs about consciousness.

What I will say is this: there are legitimate scientific questions about the relationship between consciousness and reality at the quantum level. The observer effect in quantum mechanics shows that the act of measurement changes the behavior of subatomic particles. Whether this means consciousness plays a fundamental role in creating reality or just interacts with it in specific ways remains an open question among physicists.

For our purposes, what matters is this undeniable fact: consciousness clearly determines which aspects of reality you notice, how you interpret them, and what actions you take based on those interpretations. Whether or not you're literally creating reality from quantum possibilities, you're certainly creating your experienced reality from the infinite possibilities available in any moment.

Every situation contains multiple possible interpretations and responses. Your consciousness selects which possibilities to notice and actualize through attention and action. Change your consciousness, and you change which possibilities you access.

This selective attention and interpretation is powerful enough to explain the transformation we're talking about, without needing to invoke controversial interpretations of quantum mechanics. Your beliefs determine what you notice, what it means to you, how you respond, and therefore what results you create. Change your beliefs, and you change your results, predictably and repeatably.

Complex Systems and Emergence

Your consciousness operates as what scientists call a "complex adaptive system," a network of interconnected parts that can reorganize itself based on feedback from its environment. Understanding this helps explain why small programming changes can create massive life transformations.

Complex systems have several key characteristics that are crucial for reality hacking:

Small changes can create large effects through what scientists call "sensitive dependence on initial conditions," popularly known as the butterfly effect. A tiny shift in one part of the system can cascade through interconnected feedback loops, leading to dramatic changes in overall behavior.

The whole system exhibits properties that don't exist in any individual part, which scientists call "emergence." Consciousness, for example, emerges from billions of individual neurons, none of which are themselves conscious. Similarly, your life outcomes emerge from thousands of beliefs, habits, and patterns that work together as a system.

The system adapts and evolves based on environmental feedback. When you change a belief or behavior, your whole system reorganizes to accommodate and support that change, if you maintain the new pattern long enough for the system to stabilize around it.

Feedback loops can create exponential growth or rapid collapse. Positive feedback loops amplify changes, success builds on success, confidence enables more confident action, and recognition attracts more recognition. Negative feedback loops work the same way in reverse. Failure expectations create failure behaviors, which create failure results, which reinforce failure expectations.

This systems perspective explains why some people seem to suddenly "level up" across multiple life areas simultaneously. They're not working harder on everything at once; they've changed a core piece of programming that affects the entire system, creating a cascade of improvements across all interconnected areas.

It also explains why partial changes often don't stick. If you try to change one behavior without addressing the underlying beliefs and systems that generate it, your system's natural tendency toward stability (what scientists call "homeostasis") pulls you back to your original programming.

Effective reality hacking works at the system level, changing core programming that affects multiple areas simultaneously, then reinforcing those changes until the entire system reorganizes around them.

Your Personal Reality Simulation

So, whether we're living in a literal universal simulation or not, and for practical purposes, it doesn't matter; your personal experience operates exactly like sophisticated programmable software with these characteristics:

You have a user interface (your conscious awareness) that shows you a simplified, filtered version of reality rather than overwhelming you with raw data. This interface operates according to rules determined by your programming.

You have background processes (habits, beliefs, automatic patterns) running continuously below conscious awareness, generating most of your behaviors and experiences without conscious input.

You have core programming (fundamental beliefs about reality, safety, self, and possibility) that determines how all other programs run and what results they generate.

You have subroutines (specific patterns for relationships, money, health, etc.) that execute automatically in different contexts.

You can access different processing modes (normal consciousness, flow states, glitch states) where different rules apply, and different capabilities become available.

Most importantly: you have admin access. You're not just a user of this system; you're the programmer with the authority to modify the code at any level.

The rest of this book will teach you how to exercise that authority, how to see your current programming, debug limiting code, install upgraded programs, and ultimately become the conscious master programmer of your own reality simulation.

But first, you need to learn to see the code that's currently running. That's what the rest of this chapter and the next are about: developing root access awareness, the ability to observe your own programming as it executes rather than just experiencing its outputs.

The Reality Code Audit

Before you can hack your reality, you need to understand what code is currently running. Most people are completely unconscious of their programming. To them, their automatic thoughts, emotional reactions, and behavioral patterns just feel like "reality" or "who I am."

The first skill in reality hacking is learning to observe your programming as code rather than experiencing it as reality. This is surprisingly difficult because your programming has been running automatically for so long that it feels like the truth rather than just one possible interpretation installed at some point in your past.

Here's how to begin developing this awareness. For the next seven days, spend ten minutes each evening completing this audit. Don't try to change anything yet, just observe and document.

Daily Reality Code Audit:

Automatic Thoughts: What did you think today without consciously choosing to think it? These are programs running automatically. Write down at least five automatic thoughts you noticed.

Examples might include: "I can't afford that." "They probably won't like me." "I'll fail if I try that." "This always happens to me." "I'm not smart enough for that."

Emotional Triggers: What situations created predictable emotional responses? These reveal programming about what different situations mean to you. Write down at least three emotional triggers and your automatic responses.

Examples: Speaking up in meetings leads to anxiety. Others' success leads to jealousy or inadequacy. Someone not texting back leads to rejection anxiety. Mistakes lead to shame. Unexpected changes lead to panic.

Behavioral Loops: What did you do automatically in certain situations without conscious choice? These are hardwired response patterns. Write down at least three automatic behaviors.

Examples: Procrastinating on important projects. Saying yes when you wanted to say no. Checking email compulsively. Avoiding challenging conversations. Making yourself smaller in groups.

Result Patterns: What outcomes keep repeating in your life? These are the outputs your programming generates. Write down at least three recurring patterns in your results.

Examples: Always dating unavailable people. Staying at roughly the same income level despite raises. Having the same arguments with family members. Starting projects but not finishing. Getting close to goals, then self-sabotaging.

After seven days of this audit, review all your entries and look for patterns. What themes emerge? What core beliefs might be generating these patterns? What programming seems to be running your major life areas?

Don't judge what you find. Your current programming was developed for good reasons, usually to protect you in past circumstances. The goal right now isn't to fix anything, just to see the code clearly.

Most people skip this step because it's not as exciting as learning new techniques. Don't make that mistake. You cannot hack code you cannot see. Spend the full seven days developing awareness of your current programming.

Next week, we'll learn to identify your Reality Operating System, the core programming that runs everything else. But first, you need to see the specific programs currently executing in your system.

Your homework for this week: Complete the Daily Reality Code Audit every evening. No excuses. This is the foundation everything else builds on.

The simulation has been running your entire life. For the first time, you're about to see how.

Chapter 2: Your Personal Reality Operating System

Just like your computer runs on an operating system that determines how it processes information, manages resources, and interacts with applications, your personal reality runs on what I call a "Reality OS," a core set of beliefs and assumptions that governs everything about how you experience life.

Most people are running outdated Reality OS versions that were installed during childhood and adolescence, designed for survival in environments and circumstances that no longer exist. These legacy systems continue to run automatically, generating experiences of limitation, struggle, and missed opportunities, not because the world is actually limiting them, but because their programming filters reality through an outdated lens.

I discovered this during my own transformation when I realized that my persistent struggles weren't about lack of effort or intelligence. I was working harder than most people I knew, yet getting worse results. The problem wasn't me; it was that I was running Survival OS version 1.0, programming designed for navigating childhood chaos and instability, while trying to build an adult life that required completely different code.

Understanding your current Reality OS and learning to upgrade it represents one of the most powerful reality hacks available. When you upgrade from basic survival programming to advanced creation programming, everything changes, not because you're working harder, but because you're running code that's actually designed for success rather than just avoiding catastrophe.

The Evolution of Reality Operating Systems

Most people progress through several standard Reality OS versions as they grow up, with each version optimized for different life stages and challenges. The problem is that many people never upgrade beyond the systems installed during childhood and adolescence, spending their entire adult lives running programming designed for completely different circumstances.

Survival OS 1.0 is typically installed between birth and age seven, during the period when your primary concern is physical and emotional survival. If you grew up in chaos, instability, poverty, or trauma, this becomes your foundational operating system, the deepest layer of code that influences everything else.

I remember the exact moment Survival OS finished installing in my consciousness. I was six years old, watching my father rage while my mother hid in the bathroom. Something inside me made a decision: "The world is dangerous. People can't be trusted. Safety means staying invisible. Never need anything. Never depend on anyone."

These weren't conscious thoughts; they were code being written directly into my operating system, below the level of language or choice, and they worked perfectly for their intended purpose: helping a child survive in an unstable environment. The problem is that code designed for surviving chaos at age six actively prevents you from thriving as an adult at age thirty.

Survival OS runs on several core beliefs that made sense in unstable childhood environments but create massive limitations in adult life. "The world is dangerous and unpredictable" programs your brain to constantly scan for threats, burning enormous mental and emotional energy on hypervigilance in safe environments. "Resources are scarce and must be hoarded" creates anxiety around spending money even when you have plenty, preventing investments that could multiply wealth. "Don't take risks, safety first always" blocks opportunities for growth and advancement because any uncertainty feels threatening. "Trust no one completely" prevents the deep relationships necessary for personal and professional success. "Change is threatening" makes you cling to familiar situations even when they're limiting.

Maria's story illustrates how Survival OS operates. At thirty-four, she was earning a solid salary as an accountant but living in a tiny apartment far below what she could afford, keeping six months of expenses in cash hidden in her apartment because she didn't trust banks. She'd turn down promotions that involved travel because "you never know what could happen away from home." She avoided relationships because "everyone leaves eventually." She was chronically anxious and exhausted from the constant vigilance her system maintained, even though she lived in a safe neighborhood in a stable city.

When we started working together, Maria resisted the idea that her poverty mindset and isolation were programming rather than reality. "I'm just being smart," she insisted. "The world is unpredictable. Bad things happen. I'm protecting myself."

But as we traced her beliefs back to their origins, a chaotic childhood with an alcoholic father and frequent moves, sometimes living in their car, she began to see how code written for surviving that environment was now preventing her from thriving as an adult. Her hypervigilance and hoarding made perfect sense at age eight. At thirty-four, they were creating unnecessary suffering and limitation.

The breakthrough came when she recognized that her Survival OS was still executing programs designed for a reality that no longer existed. She wasn't protecting herself; she was running outdated code that filtered out safety, opportunity, and connection because those things hadn't existed in her childhood world.

Social Acceptance OS 2.0 typically installs between ages eight and eighteen, optimized for navigating school hierarchies, peer relationships, and family expectations. This system prioritizes fitting in, following rules, and gaining approval from authority figures and peers.

David exemplified Social Acceptance OS. At twenty-eight, he was the perfect student, grown into the perfect employee, never causing trouble, following all the rules, and earning consistent approval from teachers and later managers. His programming told him: "Fitting in is more important than standing out." "Follow the rules, and you'll be rewarded." "Don't make waves or rock the boat." "What will people think?" "Authority figures know best." "Being different is dangerous."

This code served him well in traditional academic environments where success meant pleasing teachers and following prescribed paths. But it was destroying his adult potential. Despite having innovative ideas for marketing campaigns, he rarely spoke up in meetings because he worried about being seen as "too bold" or "not a team player." He stayed in an underpaid position because changing jobs felt risky and might disappoint his parents, who valued stability above all. He dated women his family approved of rather than pursuing relationships that actually excited him.

David's programming kept him perpetually seeking permission before taking action, suppressing his authentic thoughts and desires to avoid potential disapproval, following conventional paths even when they didn't align with his interests, and making decisions based on others' expectations rather than his own vision. He was successful by conventional metrics but deeply unfulfilled, living someone else's life on someone else's terms.

The wake-up call came when he was passed over for a promotion in favor of someone younger who'd been at the company half as long. His manager explained: "You're reliable and steady, but we need someone who challenges the status quo and brings bold ideas. You always wait to see what others think before committing to a position."

David was devastated, then angry. "I did everything right," he told me in our first session. "I followed the rules. I was the perfect employee. Why didn't that work?"

Because he was running Social Acceptance OS in an environment that required different programming. His code was optimized for getting approval, not for creating value or demonstrating leadership. The system that had earned him gold stars in school was now holding him back professionally.

Achievement OS 3.0 typically emerges in late teens and early twenties, focused on success within established systems through hard work, competition, and following proven paths. This is more sophisticated than earlier versions but still operates within external frameworks rather than creating its own.

Jennifer, thirty-one, represented Achievement OS at its peak efficiency and its fundamental limitations. She'd graduated summa cum laude from a top university, earned a law degree from an elite school, and worked seventy-hour weeks at a prestigious law firm, firmly on the partnership track. Her programming told her: "Hard work leads to success." "Follow the proven path, and you'll make it." "Competition is the name of the game." "Sacrifice now for rewards later." "Success is measured by external achievements."

By conventional standards, Jennifer was crushing it. By every meaningful standard, she was dying inside.

She worked exhausting hours, sacrificing health and relationships. She constantly competed with colleagues for limited partnership positions, creating a zero-sum mentality in which their success felt like her failure. She measured her worth entirely through billable hours and client wins, creating a hamster wheel where enough was never enough. She was succeeding brilliantly within someone else's system while losing herself completely.

The crisis came during a panic attack at her desk at 2 AM, preparing for a deposition while her marriage crumbled and her body gave her increasingly urgent signals she'd been ignoring. "This can't be what success is supposed to feel like," she thought as her chest tightened, and her vision blurred.

She was right. She was running Achievement OS, code designed to succeed within hierarchical, competitive systems through effort and persistence, but that programming couldn't ask the more important questions: Success toward what end? At what cost? In whose system? For whose version of winning?

Achievement OS can generate impressive external results, but it can't create fulfillment, sustainability, or genuine success on your own terms. It's still fundamentally reactive, operating within frameworks others have established, grinding toward goals others have defined as important.

The Upgrade: Advanced Operating Systems

Most people never progress beyond these basic systems. They spend their adult lives running some combination of Survival, Social Acceptance, and Achievement OS, programming designed for childhood and adolescence, optimized for navigating systems others control.

The transformation begins when you recognize that more advanced operating systems exist, and that you can deliberately install them.

Abundance OS 4.0 represents a fundamental shift from managing scarcity to creating value. Instead of competing for limited resources, fighting for approval, or grinding within established systems, Abundance OS focuses on creating opportunities, building assets, and generating value for yourself and others.

This operating system runs on fundamentally different core beliefs that change everything about how you experience reality. "Opportunities are unlimited and can be created" replaces scarcity thinking with creative possibility. "Success comes from creating value for others" shifts focus from competing to contributing. "Wealth can be generated, not just redistributed," opens possibilities beyond zero-sum thinking. "Collaboration creates better outcomes than competition" transforms relationships from threats into assets. "Systems and leverage multiply individual efforts" points toward scalable success. "Everyone can win when value is created" eliminates the false choice between your success and others' success.

Marcus's transformation from struggling freelancer to thriving business owner illustrates Abundance OS in action. As a graphic designer earning $30,000 annually, Marcus ran Achievement OS, working hard, competing on price, and trying to please every client. He measured success by hours billed and projects completed, staying busy but broke.

The shift began when a mentor asked him a simple question: "What expensive problem are you solving for your clients?"

Marcus didn't have a good answer. He was providing generic design services in a crowded market, competing primarily on price. He was running effort-based code: work hard, deliver quality, keep busy. But his programming couldn't generate the question that changed everything: "What would happen if I stopped selling hours and started selling transformations?"

He began researching his small business clients and discovered a pattern: they were losing customers due to unprofessional, inconsistent branding. He developed a systematic brand-building process that consistently increased client revenue by forty to sixty percent. Instead of charging $50 per hour for design work, he began charging $5,000 for a comprehensive branding transformation.

His income exploded, from $30,000 to $240,000 in eighteen months. But the real shift wasn't the money. It was the operating system upgrade. He stopped competing for hours and started creating value. He stopped thinking about effort and started thinking about results. He stopped working in someone else's system and started building his own.

This is what Abundance OS makes possible: seeing opportunities where others see obstacles, creating value where others compete for scraps, building systems where others trade time for money, collaborating where others compete, and generating wins where others assume zero-sum games.

Master Reality OS 10.0 represents the pinnacle of consciousness programming, a sophisticated system designed for maximum effectiveness, fulfillment, and positive impact. Very few people ever install this level of operating system, but those who do operate in a completely different reality than people running basic code.

Master Reality OS integrates everything from previous systems while transcending their limitations. It includes the survival instincts of 1.0 without the paranoia and scarcity. It honors relationships and social connections from 2.0 without requiring approval or conformity. It maintains the work ethic and goal focus of 3.0 without the grinding or burnout. It embraces the creation and collaboration of 4.0 while adding layers of sophistication most people never access.

I watched Dr. Sarah Chen operate from Master Reality OS during a medical conference where we both presented research. Where other speakers seemed nervous or competitive, Sarah radiated calm confidence and genuine interest in others' work. She'd built a thriving practice, launched a successful medical education company, maintained vibrant health, and somehow found time for her family and community service. When I asked how she managed everything, her answer revealed her operating system: "I don't manage everything. I design systems that handle most things automatically, which frees me to focus on high-leverage activities and meaningful relationships. I'm not working harder than others; I'm running different code."

Master Reality OS operates on beliefs that most people never fully embrace: "I am the conscious creator of my reality." "Everything is possible with proper understanding and application." "My success contributes to the elevation of human consciousness." "Challenges are opportunities for growth and contribution." "I am responsible for optimizing my impact on the world." "Continuous evolution and learning are my natural state."

This programming generates automatic capabilities that seem almost superhuman to people running basic code: real-time optimization and adjustment across all life areas, teaching and mentoring others in consciousness development, creation of systems that generate value beyond personal involvement, approaching challenges with curiosity and growth mindset rather than fear or stress, maintaining peak performance while preserving health and relationships, and contributing to solving important problems in the world.

The difference isn't talent, luck, or circumstances. It's an operating system. People running Master Reality OS aren't smarter or more gifted; they're running code that's optimized for creation rather than survival, for contribution rather than competition, for system-building rather than task-grinding.

Identifying Your Current Reality OS

Most people aren't running a single clean operating system. They're running a messy combination of different systems in different life areas, which creates internal conflicts and suboptimal performance. You might run Achievement OS at work while running Survival OS with money and Social Acceptance OS in relationships, creating constant friction as these incompatible programs try to execute simultaneously.

Over the next week, I want you to identify specifically what programming is running in each major area of your life. This isn't about judgment; it's about accurate diagnosis. You cannot upgrade code you haven't accurately identified.

Take out a notebook and answer these questions for each life area. Don't rush through this. Spend at least thirty minutes per area, writing your immediate responses, then diving deeper.

Money and Wealth Processing:

When you see expensive items or opportunities, what's your immediate automatic thought? Not what you think you should think, but what actually runs through your mind first? If it's "I can't afford that," you're running scarcity code. If it's "That's not for people like me," that's identity limitation code. If it's "I need to work harder to earn that," you're running effort-based programming. If it's "How could I create the value to afford that?" you're running abundance creation code. If it's "What systems could generate that level of wealth?" you're running master-level programming.

When others succeed financially, what do you feel automatically? Threatened that there's less for you reveals scarcity programming. Worried about what people think of your own finances indicates social comparison code. Motivated to work harder to compete shows Achievement OS. Inspired and curious about their strategies indicates abundance thinking. Genuinely happy for them while looking for ways to support their success suggests master-level programming.

How do you price your services or negotiate salary? Based on what you desperately need to survive indicates Survival OS. What seems socially appropriate for someone like you reveals Social Acceptance programming. Industry standards and your credentials suggests Achievement OS. The value you provide to others indicates Abundance OS thinking. Optimizing value creation for all parties involved reveals master-level code.

Opportunity Processing:

When significant opportunities appear, what's your automatic response? Assuming they're too good to be true or dangerous reveals Survival OS paranoia. Wondering if you're qualified or if others will approve shows Social Acceptance programming. Analyzing the competition and working harder to earn it indicates Achievement OS. Getting excited and looking for ways to create value suggests Abundance OS. Evaluating how to optimize the opportunity for maximum positive impact reveals master-level thinking.

When facing uncertainty, what do you typically do? Avoiding it and sticking with what's familiar indicates Survival OS. Looking for what others are doing and following their lead reveals Social Acceptance programming. Creating detailed plans and working hard to control outcomes shows Achievement OS. Seeing uncertainty as where the biggest opportunities exist indicates Abundance OS. Embracing uncertainty as a natural part of growth and evolution suggests master-level programming.

Relationship Processing:

In relationships, what's your typical pattern? Keeping people at a distance to protect yourself reveals Survival OS. Focusing on being liked and accepted shows Social Acceptance programming. Competing to be the best partner or friend indicates Achievement OS. Looking for ways to create mutual value and growth suggests Abundance OS. Seeking relationships that contribute to everyone's highest development reveals master-level code.

When conflicts arise, how do you automatically respond? Fight, flight, or freeze indicates Survival OS. Avoiding conflict to maintain harmony reveals Social Acceptance programming. Trying to win or prove you're right shows Achievement OS. Looking for win-win solutions indicates Abundance OS thinking. Using conflict as an opportunity for deeper understanding and growth suggests master-level programming.

Challenge Processing:

When facing difficult situations, what's your automatic response? Going into survival mode and just trying to get through it reveals Survival OS. Looking for what others expect you to do shows Social Acceptance programming. Working harder and pushing through with determination indicates Achievement OS. Looking for the opportunity or lesson in the challenge suggests Abundance OS. Seeing challenges as perfect opportunities for growth and contribution reveals master-level code.

When you fail or make mistakes, what do you typically feel and think? Feeling threatened and trying to avoid similar situations indicates Survival OS. Feeling shame and worrying about others' judgment reveals Social Acceptance programming. Feeling frustrated and working harder to avoid future failures shows Achievement OS. Feeling curious about what you can learn indicates Abundance OS. Feeling grateful for the growth opportunity and wisdom gained suggests master-level programming.

After completing this assessment across all areas, look for patterns. Are you running the same OS everywhere, or different systems in different areas? Where are the conflicts between incompatible programming? What areas need the most urgent upgrades?

Most people discover they're running primarily Survival and Social Acceptance OS with some Achievement OS in their career, a combination that creates constant internal conflict and suboptimal results across all areas. Understanding this clearly is the first step toward upgrading.

Sarah's Complete OS Upgrade

Let me show you what a complete Reality OS upgrade looks like by walking you through Sarah's transformation in detail. Her story illustrates not just what's possible but specifically how to implement an upgrade systematically.

Sarah, twenty-nine, worked as a marketing manager at a mid-sized company, earning $52,000 annually. She was good at her job but felt stuck, undervalued, and increasingly frustrated. Despite solid performance, she never got promoted. Despite valuable ideas, she rarely spoke up. Despite decent income, she felt perpetually broke and anxious about money.

When she came to me, frustrated and confused about why her life wasn't working despite "doing everything right," I asked her to complete the Reality OS assessment. Her results revealed the problem immediately:

At work: Social Acceptance OS 2.0 (seeking approval, avoiding standing out, following others' lead)

With money: Survival OS 1.0 (scarcity thinking, hoarding, anxiety about spending)

In relationships: Social Acceptance OS 2.0 (people-pleasing, avoiding conflict, staying in comfortable but limiting relationships)

With opportunities: Survival OS 1.0 (assuming risks are threats, staying safe, avoiding uncertainty)

She was running programming designed for surviving childhood instability and fitting into school hierarchies, trying to build an adult life that required completely different code. No wonder she felt stuck.

Week 1: System Analysis

I had Sarah spend the first week documenting her current programming in detail, without trying to change anything yet. Every evening, she completed a comprehensive code audit, writing down her automatic thoughts, emotional reactions, behavioral patterns, and result patterns across all life areas.

The patterns that emerged were striking. She realized she'd been operating on autopilot for years, running code installed decades ago, never questioning whether it still served her. Her beliefs about money were her parents' beliefs from their Depression-era childhoods. Her relationship patterns were copied from her mother's people-pleasing. Her career behavior was optimized for getting teacher approval, not for demonstrating leadership.

By the end of week one, Sarah was seeing her programming as code rather than reality for the first time. "This isn't who I am," she said with a mixture of relief and anger. "This is just programming I accepted when I was too young to know I had a choice."

That recognition, that shift from "this is reality" to "this is just code," is the crucial first step that makes everything else possible.

Week 2: Upgrade Planning

Week two focused on designing her upgraded operating system. We weren't trying to install everything at once, that's a recipe for system crashes and reversion to old programming. Instead, we identified the core beliefs that, if upgraded, would create cascade effects across all areas.

Her current core programming included: "Don't rock the boat or people won't like you." "Work hard and follow the rules, and you'll be rewarded." "Standing out is dangerous and selfish." "What will people think if I fail?" "I'm not qualified for big opportunities." "Good girls don't ask for too much."

Her new core programming would be: "My value comes from contribution, not approval." "I can be authentic and successful simultaneously." "Standing out allows me to serve others better." "My success creates opportunities for others." "I am qualified by my commitment to growth and service." "Asking for fair compensation models healthy boundaries."

The key was creating bridge beliefs, intermediate programming that felt authentic while moving her toward her desired system. You can't jump directly from "I'm not qualified" to "I'm a world-class expert" without creating cognitive dissonance that crashes the whole upgrade. But you can go from "I'm not qualified" to "I'm qualified by my commitment to learning" to "I'm developing genuine expertise" to "I'm recognized as an expert in my field."

Weeks 3-4: Core System Installation

Sarah began installing her upgraded core beliefs through a daily practice I designed specifically for her system. Every morning for fifteen minutes, she would:

Visualize herself operating from her new programming, speaking up confidently in meetings, pricing her services based on value rather than fear, being in relationships that energized rather than constrained her. She made this as detailed and emotionally compelling as possible, feeling what it would feel like to already have these upgrades installed.

Practice affirmations she'd designed herself that resonated emotionally: "I provide significant value and deserve to be seen and heard." "My authentic voice contributes to better outcomes for everyone." "I am becoming someone who creates opportunities rather than just responding to them." "My success models possibility for others."

Commit to three specific daily actions aligned with her new programming: speak up at least once in every meeting with a valuable contribution, make at least one decision based on her authentic desires rather than others' expectations, do something each day that created value for others without seeking approval.

The key was consistency. Sarah did this every single morning without exception, even when she didn't feel like it, even when it felt fake or forced. She was installing new neural pathways, and that requires repetition.

Within two weeks, she noticed significant changes. She felt more energetic and confident. Colleagues began asking for her input more frequently. She had ideas for improving processes that she'd never noticed before. Her anxiety about others' opinions decreased noticeably. She began feeling excited about possibilities rather than worried about problems.

This is what successful core system installation looks like, not dramatic overnight transformation, but steady, measurable improvements that indicate the new programming is taking hold.

Weeks 5-8: Application Installation
With her core system stabilized, Sarah systematically upgraded specific life areas:

Week 5: Financial Programming
Old code: "Don't ask for too much, be grateful for what you get."
New code: "I provide significant value and deserve fair compensation."
Implementation: She researched market rates for senior marketing positions and discovered she was underpaid by 35%. She prepared a detailed case for promotion including metrics demonstrating her value. She also raised her freelance rates by 40% and started positioning herself for higher-value clients.

Week 6: Relationship Programming
Old code: "Keep everyone happy, avoid conflict at all costs."
New code: "Authentic relationships require honest communication."
Implementation: She had a difficult but necessary conversation with her boyfriend about their future and her needs. She started attending networking events focused on meeting growth-oriented people rather than just comfortable social gatherings. She practiced saying no to requests that didn't align with her priorities.

Week 7: Career Programming

Old code: "Keep your head down and work hard, don't draw attention."

New code: "Leadership means serving others with my unique strengths."

Implementation: She volunteered to lead a cross-functional project that had been stuck for months. She started sharing strategic insights in leadership meetings. She reached out to her network about opportunities at companies whose missions excited her.

Week 8: Health Programming

Old code: "Self-care is selfish when others need me."

New code: "Taking care of myself allows me to serve others better."

Implementation: She established firm boundaries around work hours, leaving the office by 6 PM. She started a regular exercise routine, treating it as non-negotiable as any work meeting. She began meal-planning and sleep hygiene practices that supported her energy.

Each week, she focused intensely on one area while maintaining the daily core system practice. This prevented overwhelm while ensuring systematic coverage of all major life domains.

Weeks 9-12: Integration and Optimization

The final phase focused on making everything work together smoothly and handling the inevitable challenges that arise when you upgrade your programming while people around you are still running old expectations.

Integration challenges emerged quickly. Some colleagues initially resisted her more assertive communication style, interpreting her confidence as arrogance. Her boyfriend became uncomfortable with her increased independence and focus on personal growth, eventually admitting he preferred her when she was more accommodating and less challenging. Her family questioned her "selfish" focus on personal development, wondering why she wasn't as available for their needs anymore.

Sarah learned to run what I call "compatibility mode," maintaining upgraded programming internally while translating it into language others could understand. When colleagues pushed back on her directness, she explained: "I'm trying to contribute more effectively. What I'm sharing comes from genuinely wanting the best outcomes for our team." When her family questioned her boundaries, she said: "Taking better care of myself makes me a better daughter, sister, and friend. I'm not loving you less, I'm learning to love myself too."

The most difficult challenge was her relationship. Despite attempts to bridge the gap, it became clear that her boyfriend was invested in her running old programming because it served his needs. The relationship ended, which was painful but necessary. Her upgraded programming required relationships with people who supported her growth, not people who preferred her limited.

By week twelve, Sarah's entire system had reorganized around her upgraded programming. She'd been promoted to Senior Marketing Manager with a 35% salary increase. She'd launched a freelance consultancy on the side that was already generating $2,000 monthly. She'd ended her limiting relationship and was dating someone who celebrated her growth. She'd developed authentic friendships with other growth-oriented people. Most importantly, she felt like herself for the first time in years, confident, purposeful, energized, and free.

The transformation wasn't magic. It was systematic programming upgrade, consistently implemented over twelve weeks, resulting in complete system reorganization.

Common OS Upgrade Issues and Solutions

Sarah's upgrade went relatively smoothly because she followed the protocol systematically and had support. But not everyone's upgrade is that clean. Let me walk you through the most common issues and exactly how to handle them.

Issue 1: New Programming Feels Fake

This is the most common problem. You start practicing new beliefs and thoughts, but they feel inauthentic, forced, like you're lying to yourself. Your system rejects them as incompatible.

This happens when you try to jump too far too fast, installing programming that's too different from your current code without adequate bridge beliefs. It's like trying to install Windows software on a Mac without compatibility software, the system just rejects it.

Michael experienced this acutely when trying to shift from "I'm not qualified for leadership" to "I'm a natural leader." Every time he tried to think the new thought, his system immediately countered with evidence of times he'd failed to lead, mistakes he'd made, reasons why the new belief was absurd.

The solution isn't trying harder to believe something that feels false. The solution is creating bridge beliefs that feel authentic while moving you in the right direction.

Michael's bridge sequence looked like this:

Current: "I'm not qualified for leadership."

Bridge 1: "I'm learning what good leadership looks like."

Bridge 2: "I can develop leadership skills through practice and feedback."

Bridge 3: "I'm becoming more capable of leading in specific contexts."

Bridge 4: "I'm recognized as a leader by people I've helped and guided."

Bridge 5: "I'm a leader who's still growing and developing."

Final: "I'm a natural leader" (now feels true because of the journey)

The key is that each step feels like a truthful statement of where you currently are or where you're clearly headed. You're not lying to yourself, you're accurately describing your trajectory. This allows the new programming to install smoothly rather than being rejected as incompatible.

Spend a week on each bridge belief before moving to the next. Install it through daily visualization and practice, take actions that support it, collect evidence that confirms it, then progress to the next level. This gradual installation prevents system crashes and rejection.

Issue 2: Old Programming Resurfaces Under Stress

This is frustratingly common and can make you feel like you've made no progress. Your new programming works perfectly in normal situations, but the moment stress hits, you revert completely to old patterns, the anxiety, the people-pleasing, the scarcity thinking, all of it comes flooding back.

Jennifer experienced this intensely. She'd done great work upgrading her programming around self-worth and confidence. In normal situations, she felt genuinely confident and capable. But the moment her demanding CEO called an unexpected meeting, her old people-pleasing and anxiety programming took over completely. She'd spend hours preparing, lose sleep worrying, and show up to the meeting apologizing before she even spoke.

This happens because stress activates your oldest, most established neural pathways, your "default" programming. Under pressure, your brain automatically runs the code that's been in place longest because it feels safest and most familiar. Your new programming hasn't been tested under fire yet, so your system doesn't trust it in high-stakes situations.

The solution is creating specific protocols for high-stress situations and deliberately practicing your new programming under gradually increasing pressure.

Jennifer created a "CEO Interaction Protocol" that activated automatically before these high-stress meetings:

Ten minutes before the meeting, she'd visualize handling it from her upgraded programming, speaking confidently, contributing value, remaining calm regardless of his reaction.

She created a physical anchor, touching her watch, that reminded her body and mind: "I am operating from upgraded programming now."

She practiced a breathing protocol, three deep breaths, before speaking, which gave her nervous system time to regulate and her new programming time to load.

She repeated an internal phrase, "I am here to create value," that oriented her toward contribution rather than approval-seeking.

After the meeting, she'd document what worked and what needed adjustment, treating each interaction as practice for the next.

Within three meetings using this protocol, her new programming became automatic even in high-stress CEO interactions. The key was deliberate practice under pressure, not just hoping the new code would work when it mattered.

Issue 3: Others Resist Your Upgrade

This catches most people by surprise. You'd think people who care about you would celebrate your growth and support your changes. Sometimes they do. Often, they don't.

When David started upgrading from Social Acceptance OS to Abundance OS, his family reacted with confusion and criticism. His mother complained he was "becoming arrogant." His father warned he was "getting too big for his britches." His longtime friends stopped inviting him to gatherings, saying he'd "changed" in a way that made them uncomfortable.

This happens for several reasons. Your changes threaten existing relationship dynamics and power structures. If you've always been the accommodating one, the helper, the one who makes everyone comfortable, people have organized their lives around that role. When you stop playing it, they feel destabilized.

Your growth makes others uncomfortable with their own limitations. When you start taking risks and growing, it highlights that they're staying stuck. Rather than feeling inspired, they often feel judged or inadequate, not because you're judging them, but because your changes force them to confront their own programming.

Some people are genuinely invested in keeping you limited because it serves their needs. They preferred you with old programming because it made their lives easier, even if it made your life harder.

The solution isn't trying to convince them you're still the same person, because you're not. You're running different code, which means you're genuinely different. The solution is what I call "compatibility mode," maintaining your upgraded programming while translating it into language that doesn't threaten others.

David learned to explain his changes in ways his family could hear: "I'm not rejecting the values you taught me, I'm building on them. You taught me about hard work and integrity. I'm applying those principles in ways that create more impact and opportunity." When friends said he'd changed, he agreed: "I have changed, and I'm grateful for it. I'm becoming who I'm meant to be. That doesn't mean I care about you less, it means I can show up as a better friend."

He also accepted that some relationships were based on his old programming and wouldn't survive the upgrade. This was painful but necessary. His job wasn't to stay limited so others felt comfortable. His job was to grow and find people who celebrated that growth rather than resented it.

Within six months, David had new friends who were excited about growth, family members who eventually accepted his changes (though some never did), and clarity about which relationships genuinely supported him versus which required him to stay small.

Issue 4: Partial Installation Creates Conflicts

This happens when you upgrade some areas while leaving others running old programming, creating internal conflicts that can be paralyzing.

Marcus experienced this when he upgraded his money programming to Abundance OS while leaving his relationship programming on Social Acceptance OS. His money code told him: "Create value, charge what you're worth, build systems." His relationship code told him: "Keep everyone happy, avoid conflict, don't appear greedy."

The conflict became acute when he needed to raise rates on existing clients. His money programming said this was necessary and appropriate, his value had increased, his services commanded higher prices, clients were getting better results. His relationship programming triggered panic: "They'll think you're greedy. They'll be angry. They'll leave. You'll lose their approval."

He'd try to have the conversation about rates, then immediately backpedal to keep everyone happy. He'd start building boundaries around his time, then cave when clients pushed back. He was running incompatible programs simultaneously, creating internal paralysis and external confusion.

The solution was systematically upgrading all related systems so they operated coherently. Marcus couldn't run Abundance OS for money and Social Acceptance OS for relationships, those programs conflicted directly. He needed to upgrade his relationship programming to be compatible with his upgraded money code.

His new relationship programming became: "I create value for clients through clear boundaries and appropriate pricing." "Respecting myself and my worth models healthy relationships." "People who value me will respect my growth and boundaries." "Those who don't aren't right clients for me."

This created consistency across systems. His money programming and relationship programming now worked together instead of fighting each other. He could raise rates, set boundaries, and select clients aligned with his values because all his systems supported those actions.

Issue 5: Progress Plateaus

This is demoralizing. You make great initial progress, new beliefs installing, behaviors changing, results improving. Then suddenly everything stops. No more progress. Sometimes even regression. You feel stuck at an intermediate level, unable to break through to the next stage.

Lisa experienced this intensely during her upgrade. Weeks three through eight showed steady, dramatic improvements. She felt like a different person. Then weeks nine through twelve showed almost no additional progress. She was better than before but stuck, unable to reach the vision she'd created for herself.

This happens because your system reaches a temporary equilibrium, a local maximum where your programming is better than before but not yet fully upgraded. Your system stabilizes at this intermediate level and resists further change, trying to establish new patterns as permanent rather than continuing to evolve.

The solution is deliberately disrupting this equilibrium through what I call "advancement challenges," specific actions that are just beyond your current capability, forcing your system to reorganize at a higher level.

For Lisa, this meant choosing increasingly challenging implementations of her new programming. During the plateau, she'd been practicing her upgraded code in comfortable situations where it worked easily. To break through, she needed to test it in situations that stretched her capabilities.

She volunteered to lead a major presentation to senior leadership, something that would have terrified her old programming and was still uncomfortable for her new programming. She started a podcast sharing her expertise, putting herself visible in a way that challenged both old and new programming. She applied for a position two levels above her current one, testing whether her upgraded code could handle that level of stretch.

Each challenge pushed her system past its comfortable equilibrium, forcing reorganization at a higher level. Within three weeks of implementing advancement challenges, her progress resumed. Six weeks later, she'd broken through the plateau completely and was operating at a level she hadn't imagined possible at the start.

The key insight: systems stabilize at new levels and resist further change unless deliberately challenged. Progress requires continuously pushing just beyond your current comfort zone, forcing ongoing evolution rather than settling into a new comfortable pattern.

Your OS Upgrade Plan

You now understand the different Reality Operating Systems, how to identify which you're currently running, and how to implement systematic upgrades. Now it's time to create your specific plan.

Take out a notebook and complete this planning process. Don't rush through it. Spend several hours over the next few days really thinking through your current programming and designing your upgrade path.

Step 1: Current System Assessment

Complete the Reality OS diagnostic for all major life areas (money, career, relationships, health, personal growth). Write down specifically what beliefs and patterns you're currently running. Don't judge or try to fix anything yet, just get an accurate diagnosis.

Most people discover they're running Survival OS 1.0 for money, Social Acceptance OS 2.0 for relationships and career, and Achievement OS 3.0 for some professional areas. This combination creates constant internal conflict and suboptimal results.

Step 2: Identify System Conflicts

Look at the programming you identified in different areas. Where do incompatible systems create conflicts? Write down specific situations where different programming creates paralysis or contradictory actions.

Example: "My money programming (Survival OS) says save everything and avoid risks. My career programming (Achievement OS) says work hard and compete. These conflict when opportunities require investment or risk-taking for advancement."

Step 3: Design Target System

What programming do you want to be running in each area? Be specific. Don't just write "better programming," define exactly what beliefs, automatic thoughts, and behavioral patterns you want installed.

Remember: aim for Abundance OS 4.0 or Master Reality OS 10.0. Don't design upgrades that just optimize your current limitations. Design programming that will actually generate the life you want.

Step 4: Create Bridge Beliefs

For each major belief upgrade, design bridge beliefs that will get you from current to target programming without creating system rejection.

Create three to five intermediate beliefs that feel authentic while moving you in the right direction.

Step 5: Design Daily Practice
Create a specific daily practice for installing your core programming. This should take ten to fifteen minutes each morning and include visualization, affirmation, and action commitment.

Your daily practice must be specific enough that you could teach someone else to do it exactly. "Think positive" is too vague. "Spend ten minutes visualizing yourself speaking confidently in meetings, focusing on how it feels in your body, then commit to contributing one valuable idea in today's team meeting" is specific enough to implement.

Step 6: Plan Weekly Implementations
Map out which specific life area you'll focus on each week for weeks five through eight. Create specific actions you'll take to implement new programming in that area.

Step 7: Identify Support Systems
Who will support your upgrade? What environments support your new programming? What resources do you need? What potential obstacles should you prepare for?

Step 8: Establish Metrics
How will you know your upgrade is working? What specific, measurable changes should you see at four weeks, eight weeks, and twelve weeks?

Don't make all your metrics external (income, promotions, etc.). Include internal metrics: How do you feel daily? What thoughts run automatically? What do you notice that you didn't before? How do you respond to challenges?

Step 9: Create Emergency Protocols
Plan for high-stress situations, system crashes, and the resurfacing of outdated programming. Create specific protocols for how you'll handle these predictable challenges.

Step 10: Schedule Review Points

Put actual calendar appointments for weekly reviews (30 minutes) and monthly comprehensive audits (2 hours). Treat these as seriously as any work meeting. Without consistent review and adjustment, upgrades often fail.

The Twelve-Week Transformation Timeline

Here's what a successful OS upgrade timeline looks like:

Week 1: Recognition and Documentation

You're seeing your current programming as code rather than reality for the first time. This feels disorienting but liberating. You're documenting patterns, beliefs, and automatic responses across all life areas.

Week 2: Design and Preparation

You're designing your upgraded operating system and creating bridge beliefs. This feels exciting and slightly overwhelming. You're preparing your daily practice and implementation plans.

Weeks 3-4: Core System Installation

You're practicing daily visualization, affirmation, and action. This feels awkward and sometimes fake at first, then gradually more natural. You're seeing small changes, more confidence, better energy, different thought patterns.

Weeks 5-8: Application Installation

You're systematically upgrading specific life areas one per week. This feels intense but manageable. You're seeing real results, better conversations, improved opportunities, increased income, deeper relationships.

Weeks 9-10: Integration

You're handling conflicts between upgraded and old programming, resistance from others, and system stability issues. This feels challenging but navigable because you have protocols.

Weeks 11-12: Optimization

You're fine-tuning, handling advanced challenges, and establishing maintenance routines. Your upgraded programming is starting to feel normal, like this is just who you are now.

The timeline isn't magic, it's based on how long consistent neural rewiring typically takes. Some people move faster, some slower, but twelve weeks of consistent practice creates measurable, stable transformation for most people.

Maintenance: Keeping Your Upgraded OS Running

The biggest mistake people make after successful upgrades is assuming the work is done. Your upgraded programming needs ongoing maintenance to stay stable and continue evolving.

I learned this the hard way during my second year of nursing school. I'd successfully upgraded from Survival OS to Achievement OS, transforming my academic performance and confidence. Then finals week hit with unexpected intensity, multiple exams, family crisis, relationship problems, health issue, all simultaneously.

Under that extreme stress, my system crashed completely. I reverted to pure Survival OS overnight, panic, anxiety, scarcity thinking, hypervigilance, all my old patterns flooding back like they'd never left. I felt devastated. All that work, and I was right back where I started.

My mentor helped me understand what happened. "Your upgraded programming is installed but not yet deeply rooted," she explained. "It's like a newly planted tree. It looks strong, but it needs consistent care until its root system is established. Extreme stress knocked your system back to its oldest, deepest programming. That doesn't mean you lost your progress, it means you need better maintenance protocols."

She was right. Once I implemented consistent maintenance practices, my upgraded programming stabilized and eventually became my default even under pressure. Now I can handle extreme stress while maintaining my upgraded systems because those systems have deep roots.

Here's the maintenance protocol that keeps upgraded programming stable:

Daily: Morning System Check (5 minutes)

Brief visualization of yourself operating from upgraded programming today. Quick check: What programming am I running right now? Any old patterns trying to resurface? What's my energy level and mental state? What's my intention for today?

Daily: Evening Review (5 minutes)

Review how well you operated from upgraded programming today. Celebrate successful executions, this reinforces new neural pathways. Note when old programming ran, this helps you identify triggers. Plan tomorrow's focus based on today's insights.

Weekly: Comprehensive Review (30 minutes)
Assess the week's programming execution across all areas. Analyze any patterns in when old programming resurfaces. Identify what's working well and what needs adjustment. Plan next week's focus and implementations. Check that your support systems are solid.

Monthly: Deep Audit (2 hours)
Complete system review of all programming and results. Assessment of progress toward major goals and vision. Identification of areas needing additional work or optimization. Planning for next month's focus and development. Adjustment of daily and weekly practices based on what's working.

Quarterly: Strategic Planning (half day)
Major assessment of your entire operating system. Evaluation of progress over past three months. Design of next-level upgrades and challenges. Network and support system optimization. Integration of major learnings and breakthroughs.

Annually: Vision and Evolution (full day)
Comprehensive review of the entire year's progress. Celebration of growth and transformation. Assessment of where you are versus where you intended to be. Design of major upgrades for the coming year. Vision clarification and strategic planning.

This might sound like a lot of time, but consider: you spend hours daily on social media, entertainment, and activities that don't improve your life. Investing thirty to sixty minutes daily in maintaining and optimizing your reality operating system is the highest-leverage use of time possible.

A Final Word on Operating Systems
Your Reality OS is always running, always executing, always generating your experience. The question has never been whether you're running an operating system, you are, and it's generating your current results. The only question is whether you'll continue running whatever code was installed

automatically during childhood and adolescence, or whether you'll take conscious control and upgrade to programming designed for the life you actually want.

Most people never even consider that their operating system can be upgraded. They assume their limitations, struggles, and stuck patterns are just "reality" or "who I am." They work harder, try different strategies, pursue new opportunities, all while running the same underlying code that created their limitations in the first place.

It's like trying to run advanced software on an outdated operating system. No matter how hard you work, no matter what applications you install, if your core OS is designed for survival rather than creation, you'll keep hitting the same limitations.

But you're different. You're reading this book, which means some part of you already knows your current programming is optional. You're ready to see your limitations as code rather than reality. You're ready to become the conscious programmer of your own operating system.

The simulation has been running your entire life on inherited code. It's time to install your own programming.

Chapter 3: Finding Reality Glitches

The first time you experience a genuine glitch in your reality simulation, you'll know it beyond any doubt. Your normal sense of limitation will temporarily vanish, replaced by an almost supernatural feeling of expanded possibility. Time will shift in ways that defy logic, three hours evaporating like thirty minutes, or a single moment stretching into what feels like an eternity of clarity. Your usual anxieties and self-doubts will simply disappear, not because you've suppressed them but because they'll feel irrelevant, like someone else's problems from another lifetime.

In that state, problems that seemed impossibly complex will suddenly appear simple, almost laughably so. Solutions will emerge fully formed in your consciousness as if they'd been waiting just beneath the surface of your normal awareness. You'll wonder how you never saw them before, how something so obvious remained invisible for so long.

Most people experience these moments and dismiss them. They call them random inspiration, lucky breaks, or temporary delusions. They have a breakthrough insight during a long run, feel invincible after a powerful conversation, or suddenly see their life with perfect clarity during a moment of crisis, then return to normal consciousness without recognizing what just happened or how to access it again.

This is one of the great tragedies of the human experience: we regularly access glitch states where rapid transformation becomes possible, then dismiss these experiences as flukes rather than recognizing them as windows into the programmable nature of reality itself.

I'm going to teach you to recognize these glitch states when they occur, understand what creates them, learn to trigger them deliberately, and most importantly, exploit them to install permanent upgrades to your reality programming. This might be the most valuable skill you learn from this entire book, because glitch states provide the cleanest, fastest path to transformation.

What Makes a Glitch State?

During my first year of nursing school, I experienced a glitch so profound it changed not just my academic trajectory but my entire understanding of human potential. I've already mentioned how I was struggling pathophysiology despite spending more hours studying than almost anyone in my class. What I haven't told you is what that glitch state actually felt like from the inside, because understanding its phenomenology, the subjective experience, is crucial for learning to recognize these states in your own life.

It was three-seventeen in the morning. I remember checking the clock, feeling that sick desperation that comes from knowing you're about to fail something that matters desperately. The final exam was in six hours. I'd been studying since six PM the previous evening, nearly nine hours of grinding through material that felt like it was written in a language my brain simply couldn't process.

Then something shifted. I can't tell you exactly what triggered it, maybe pure exhaustion pushed my brain past some threshold, maybe desperation somehow suspended my normal limiting patterns, maybe the universe decided to throw me a bone. What I can tell you is that one moment I was drowning in confusion and panic, and the next moment I was experiencing reality completely differently.

The first thing I noticed was the silence. My usual mental chatter, the constant stream of worry, self-criticism, and anxiety that had been my companion through every study session, simply stopped. Not like I was suppressing it or fighting it, but like someone had turned off a radio that had been playing static my entire life. In the silence that remained, I could actually think clearly for the first time in months.

The molecular structures in my textbook, which had looked like meaningless abstract symbols moments before, suddenly resolved into something I can only describe as beautiful. I could see relationships between compounds the way you might suddenly see a hidden image in one of those 3D stereogram pictures, it was always there, but your perception has to shift in a specific way before you can see it.

But the most profound change was in my sense of self. I wasn't "Sarah, the girl from a chaotic background who's not smart enough for nursing school and is about to prove everyone right about her limitations." I was just... consciousness engaging with information. The identity I'd been carrying, with all its limitations and fears and historical baggage, felt like a costume I'd been wearing without realizing it, and in that moment, I'd somehow stepped out of it.

Time did something strange. I studied intensively for six more hours, but it felt like maybe thirty minutes had passed. Not because I wasn't paying attention to time, but because I was so completely absorbed that time stopped feeling relevant. When my alarm went off at nine AM for the exam, I looked at the clock in genuine surprise. How could six hours have passed?

More importantly, during those hours, complex concepts that had been completely opaque suddenly made perfect sense. I wasn't working harder or using different study techniques, my brain was processing the same information in a completely different way. It was like someone had upgraded my cognitive operating system overnight.

I walked into that exam feeling calm, confident, and genuinely curious about the questions. Not because I'd magically memorized everything, but because I understood the underlying principles in a way that made specific memorization unnecessary. I scored in the top ten percent of the class.

But here's the crucial part that most people miss when they hear transformation stories like this: The glitch state didn't give me capabilities I didn't have. It removed the programming that was blocking me from accessing capabilities I'd always had.

My brain was always capable of understanding pathophysiology. My intelligence was always sufficient. The limitation was code, programming installed through years of struggle and failure in chaotic circumstances, beliefs about what someone from my background could achieve, identity constructs about who I was and what I was capable of. When that programming temporarily suspended during the glitch state, my natural capabilities could finally express themselves.

This is the profound secret of glitch states: they don't create magic. They reveal that what you thought was fixed reality is actually just programming that can be suspended and rewritten.

The Neuroscience of Glitch States

From a scientific perspective, glitch states involve what researchers call "transient hypofrontality," a temporary reduction in the prefrontal cortex's normal control and inhibition functions. Dr. Arne Dietrich's research at the

American University of Beirut shows that during these states, your brain operates fundamentally differently than during normal consciousness.

Your prefrontal cortex, the part of your brain responsible for executive function, self-criticism, and filtering, essentially quiets down. This region normally acts like a security system, constantly checking everything against your existing beliefs and patterns, rejecting information and responses that don't match your programmed identity and capabilities.

During glitch states, this security system temporarily goes offline. Without your prefrontal cortex's constant interference, other brain regions can communicate in novel ways. Your default mode network, the system responsible for self-referential thinking and your sense of "I," also quiets down, which explains why your sense of identity often becomes fluid or expansive during these states.

Simultaneously, your brain chemistry shifts. Dopamine and norepinephrine levels often increase, creating heightened attention and motivation. In some glitch states, serotonin or endorphins increase, creating feelings of wellbeing and connection. These neurochemical changes create windows where your brain becomes more plastic, more receptive to new patterns and less bound by existing ones.

Brain imaging studies of people in flow states, meditation, during peak experiences, or under certain psychoactive substances show remarkably similar patterns: reduced prefrontal activity, quieter default mode network, altered brain wave states, and increased communication between brain regions that don't normally talk to each other.

Think of it like this: normal consciousness is like your computer running multiple security programs, virus scanners, and firewalls simultaneously. Everything gets checked against existing rules before it's allowed to execute. Glitch states are like temporarily putting your system into safe mode, many of those security programs are suspended, allowing direct access to core programming that's usually protected.

This creates a window where you can install new beliefs that would normally be rejected as incompatible, access capabilities that are normally locked, make changes to core programming that's usually protected, and experience yourself in ways that contradict your usual identity.

Types of Reality Glitches

Not all glitch states are the same. Understanding the different types helps you recognize them when they occur and learn to trigger specific types deliberately.

Perception Glitches are moments when you suddenly see familiar situations through completely different lenses, as if someone switched the filter on your reality. The situation hasn't changed, but your interpretation of it transforms so dramatically that it might as well be a different situation.

Rebecca experienced a powerful perception glitch during what should have been one of the most frustrating moments of her week. She was stuck in traffic on the interstate, late for an important meeting, with construction turning a normally twenty-minute drive into what was approaching an hour. She could feel her usual anxiety building, the tightness in her chest, the irritated internal monologue about incompetent city planners and terrible drivers.

Then something shifted. In one moment, she was trapped in traffic, a victim of circumstances beyond her control, her morning ruined. In the next moment, she saw it completely differently. She wasn't trapped, she was gifted unexpected time for peaceful reflection. No one could reach her. No demands could be made. She was suspended in a metal cocoon with nothing to do but breathe and think and exist.

The traffic hadn't changed. Her meeting was still waiting. The objective circumstances were identical. But her experience of those circumstances had transformed completely. Instead of fighting reality, she relaxed into it. Instead of frantically checking her phone and stressing about what she was missing, she turned off her radio and just... breathed.

Here's what makes this a true glitch rather than just positive thinking: the new perception didn't feel like she was trying to put a positive spin on a bad situation. It felt like she was seeing clearly for the first time, and her previous interpretation had been the distortion. Her usual stress response to traffic felt like someone else's problem, a pattern she'd been running automatically without ever questioning whether it served her.

That perception shift lasted for weeks afterward. Traffic stopped being something that happened to her and became just... traffic. Neutral. Sometimes convenient, sometimes inconvenient, never personal. Twenty years of automatic stress response to a common situation, transformed in a single glitch moment.

This is what perception glitches make possible: permanent transformation of how you experience recurring situations, not through effort or forcing yourself to think differently, but through a shift in perception so fundamental that your old interpretation becomes literally impossible to access.

Capability Glitches are moments when you perform beyond your normal abilities, accessing skills or confidence you didn't know you had. Your usual limitations simply don't apply.

Marcus was terrified of public speaking to the point of turning down promotions that would require presentations. He'd had a traumatic experience in high school, freezing during a presentation while classmates laughed, that installed deep programming around public speaking being dangerous and himself being incapable.

But during a company crisis when leadership was paralyzed and no one was stepping up to communicate with panicked employees, something in Marcus shifted. He didn't think about it or plan it. He just stood up during a tense meeting and started speaking, clearly, confidently, compellingly. He articulated the problem, acknowledged people's fears, outlined a path forward, and inspired his entire team.

While he was speaking, he felt no fear. His usual self-consciousness was completely absent. He wasn't thinking about how he looked or sounded or whether people were judging him. He was completely absorbed in the communication itself, in the need to help people understand and feel reassured.

When he finished, his team applauded. His manager stared at him in shock. "Marcus, I've never seen you speak like that. That was incredible. Where has this been?"

Where had it been? It had always been there, underneath programming that said public speaking was dangerous and he was incapable. The crisis created conditions that suspended that programming temporarily, allowing his natural capability to express itself.

The crucial question was whether Marcus would recognize this as a glitch showing him his actual capability or dismiss it as a one-time fluke driven by adrenaline and necessity. Most people would have dismissed it. Marcus, because he understood glitch states, recognized it for what it was: proof that his limitation was programming, not reality.

He used that glitch as evidence to rewrite his entire identity around public speaking. Every time his old anxiety programming tried to activate before a presentation, he'd remember that moment during the crisis. "That's who I actually am," he'd think. "The fear is just old code."

Within six months, Marcus was regularly volunteering for presentations and doing them well. The glitch hadn't given him magical speaking abilities, it had shown him that his perceived inability was just programming that could be overridden.

Opportunity Glitches are moments when resources, connections, or possibilities appear that "shouldn't" be available to someone in your situation, as if normal rules of access temporarily don't apply.

Jennifer graduated from college with a communications degree, no connections in her desired industry, and a resume that looked identical to ten thousand other recent graduates. The conventional wisdom said she'd need to start in an entry-level position, work for several years building experience and connections, then maybe eventually work her way into something close to what she actually wanted to do.

But during a flight to visit her parents, she struck up a conversation with the woman sitting next to her. They talked about books, travel, life philosophies, just a pleasant conversation with a stranger. Near the end of the flight, the woman asked what Jennifer did.

"I just graduated," Jennifer said. "I'm trying to break into public relations, but I don't really have connections yet. I'm applying for entry-level positions and hoping something opens up."

The woman smiled. "What kind of PR work interests you specifically?"

Jennifer described her ideal role, working with tech companies to shape their public narrative, helping innovative organizations tell compelling stories. The woman listened, asked thoughtful questions, then handed Jennifer her business card.

She was the VP of Communications for a major tech company. "Send me your resume. We have an opening that sounds exactly like what you just described. Can't promise anything, but I'd like to pass it to our hiring manager."

Three weeks later, Jennifer had the job, not an entry-level position, but the actual role she wanted, at a company she admired, with a salary that exceeded what she'd hoped for five years in the future.

Random luck? Maybe. But here's what makes it a glitch: Jennifer noticed this pattern of unexpected access repeated throughout her life. Doors opened that shouldn't have opened. People offered help that seemed disproportionate to what she'd done. Resources appeared just when she needed them.

Most people experiencing this would just call themselves lucky and leave it at that. Jennifer, once she understood glitch states, recognized a pattern. These opportunities correlated with specific internal states, moments when she was genuinely curious rather than desperate, present rather than anxious about future outcomes, offering value in conversations rather than trying to extract it.

She began deliberately cultivating those internal states, and the "lucky" opportunities increased in frequency. She wasn't creating magic, she was running programming that made her more likely to notice opportunities, more attractive to people who could help her, more open to possibilities that didn't match her preconceived ideas about how things should work.

Synchronicity Glitches are moments when external events align in statistically improbable ways to support your goals or answer your questions, as if reality is responding to your consciousness.

David needed exactly three thousand dollars for a professional certification course that could transform his career. He'd been saving for months but was still twelve hundred dollars short. Registration closed in three days. He'd run every scenario, taking on extra projects, asking for a loan, putting it on a credit card and hoping he could pay it off. Nothing felt right.

He finally decided to let it go. "If it's meant to happen, it'll happen," he thought, releasing his attachment to forcing the outcome.

The next morning, he received a notification from the IRS. Due to an error in his previous year's return, he was owed a refund. The amount: exactly three thousand dollars. Not twenty-nine hundred. Not thirty-one hundred. Exactly the amount he needed, arriving exactly when he needed it.

Coincidence? Possibly. But David noticed these kinds of "coincidences" happened with unusual frequency when he held clear intentions without desperate attachment. When he needed specific information, he'd randomly encounter an article addressing exactly that question. When he needed to connect with someone in a particular industry, that person would somehow appear in his network. When he needed resources, opportunities to generate exactly those resources would materialize.

Carl Jung called these meaningful coincidences "synchronicities," events related not by causation but by meaning. Whether they represent some mysterious property of consciousness affecting physical reality, or just heightened pattern recognition and selective attention is debatable. What's not debatable is that paying attention to these synchronicities and acting on them creates better outcomes than dismissing them as random chance.

David learned to treat synchronicities as feedback from reality. When improbable alignments occurred, he took them seriously. When doors opened unexpectedly, he walked through them. When resources materialized at the perfect time, he used them. This practice didn't require believing in mystical explanations, it just required noticing patterns and responding to them intelligently.

Flow State Glitches are moments when you enter "the zone" where time disappears, effort becomes effortless, and your performance dramatically elevates. Athletes call it being "in the zone." Writers call it "channeling." Musicians call it "feeling the music."

Lisa, a writer struggling with a difficult chapter in her novel, experienced this dramatically. She'd been working on the chapter for three weeks, writing and deleting, forcing and fighting, producing wooden prose that she knew wasn't working. Every session felt like trying to run through mud, exhausting effort for minimal progress.

Then one Tuesday morning, something shifted. She sat down at her computer without expectations, just showing up to put in her time. She wrote the first sentence, then the second. By the third paragraph, she'd entered a state she'd only experienced a handful of times before.

The character's voice became so clear it was like transcription rather than creation. She knew what they would say and do without consciously deciding. The prose flowed without effort, not because it was sloppy or unedited, but because it emerged in essentially final form. She wrote for what felt like maybe forty-five minutes.

When she looked up, four hours had passed. The difficult chapter was complete, and reading it back, it was some of the best writing she'd ever done. She'd produced more quality work in that four-hour flow session than in three weeks of forced grinding.

Most writers experience flow occasionally and treat it as a mysterious gift that strikes unpredictably. Lisa began studying what created those states. She noticed patterns: they occurred when she showed up without expectations rather than desperately needing a good session, when she was writing for the sake of the writing rather than focusing on outcomes, when she'd warmed up with easy writing before attempting difficult sections, when she eliminated all distractions and gave the work her complete attention.

She couldn't force flow, but she could create conditions that made it more likely. Within six months, she was entering flow states regularly rather than occasionally. Her productivity increased by a factor of three while her effort actually decreased.

This is the power of recognizing flow states as glitches rather than random gifts: you can learn to trigger them systematically rather than waiting for inspiration to strike.

My Pathophysiology Glitch: The Complete Story

I need to tell you the complete story of my pathophysiology glitch because understanding what happened at a deeper level will help you recognize and exploit your own glitches.

The context matters. I was twenty-two years old, first-year nursing student, carrying trauma and chaos from a childhood that would have broken many people. I'd already overcome enormous obstacles just to get into nursing school. But Pathophysiology was defeating me in a way that triggered every insecurity and limiting belief I'd ever had about my intelligence and worthiness.

I wasn't just struggling with a difficult subject. I was facing the potential collapse of my entire identity reconstruction. I'd built a fragile new sense of self as "someone capable of going into medicine," and that identity was about to be demolished by my inability to understand molecular structures.

The night before the final, I'd been studying in the library until it closed at midnight, then moved to a twenty-four-hour diner where I bought coffee I couldn't afford and tried to make sense of concepts that felt like they were written in a foreign language my brain simply couldn't parse.

By three AM, I was beyond exhausted. I'd crossed some threshold where my normal anxiety and self-criticism were too tired to keep running. My usual mental chatter, the constant stream of "you're not smart enough, you're going to fail, everyone was right about your limitations," had exhausted itself into silence.

In that silence, something extraordinary happened. Without my usual limiting programming running interference, my actual intelligence could finally engage with the material. The molecular structures I'd been staring at for weeks suddenly resolved into patterns. I could see how they related to each other, how changes in one part of a molecule would affect other parts, how different reactions followed logical principles.

But it wasn't just intellectual understanding. Something shifted at an identity level. I'd been approaching the material as "Sarah, who's not good at science, struggling with something she'll never understand." In the glitch state, that entire identity construct dissolved. I wasn't "someone bad at science" trying to force myself to understand something impossible. I was just consciousness engaging with information. The limitation simply wasn't there.

Time dilated in that strange way that happens during glitch states. I studied intensively from three AM until nine AM, six hours of continuous focused work. But experientially, it felt like maybe thirty minutes had passed. Not because I wasn't aware of time, but because I was so completely absorbed that time stopped being relevant. Past and future collapsed into an eternal present moment of engagement with fascinating patterns.

When my alarm went off at nine AM for the exam, I felt calm, energized, and genuinely curious about the questions. This might sound impossible if you've never experienced a glitch state, how can you go from panic and exhaustion to calm energy after an all-night study session?

Here's what I understand now that I didn't then: the exhaustion, panic, and overwhelm weren't coming from the studying itself. They were coming from the constant mental and emotional effort of running limiting programming while simultaneously trying to learn. The anxiety, self-doubt, and identity threat were burning enormous energy, making the actual learning far more difficult than it needed to be.

When that programming temporarily suspended, I could engage with the material directly. The studying itself wasn't actually that hard, it was just hard when filtered through programming that said it should be impossible for someone like me.

I scored in the top ten percent of the class. But more importantly, I had proof that my limitations were programming rather than reality. This single experience changed my entire trajectory because I never again fully believed in my own limitations the way I had before. When I encountered something difficult, I knew at a deep level that the difficulty came from my programming, not from the task itself.

The Anatomy of Glitch States

After experiencing that pathophysiology breakthrough and eventually learning to trigger glitch states deliberately, I've identified common characteristics that define these experiences. Understanding these characteristics helps you recognize glitches when they occur and distinguish them from other mental states.

Characteristic 1: Quiet Mind

Normal consciousness involves constant mental chatter, a stream of thoughts, worries, judgments, plans, and commentary running continuously in the background. During glitch states, this mental noise significantly quiets or stops entirely. Not through force or meditation technique, but spontaneously.

This quiet isn't empty or void, it's more like clear space where thoughts can arise without being drowned out by noise. When thoughts do come, they're clearer, more focused, more purposeful. There's no arguing with yourself, second-guessing, or running multiple contradictory mental processes simultaneously.

The first time you experience this, it's startling. You might not even realize the mental chatter was there until it stops, like suddenly noticing a humming sound you'd tuned out when it finally ceases.

Characteristic 2: Identity Fluidity

Your normal sense of self, "I am this kind of person with these capabilities and limitations," becomes flexible or temporarily dissolves. You're not experiencing yourself as your usual identity construct but as something more fundamental, awareness, consciousness, being.

This explains why limitations can suspend during glitches. The limitation isn't in your actual capabilities, it's in your identity programming. When identity becomes fluid, those programmed limitations lose their grip.

This doesn't mean you become someone else or lose yourself. It means you experience yourself as less fixed, more expansive, not bound by the stories and limitations you usually carry.

Characteristic 3: Altered Time Perception

Time becomes subjectively distorted. Hours feel like minutes, or single moments stretch into what feels like extended duration. You're not hallucinating or losing touch with reality, clocks still show objective time passing normally. But your subjective experience of time shifts dramatically.

This often correlates with being completely absorbed in activity. When you're totally present, not thinking about past or future, conventional time perception breaks down. You enter what could be called "timeless time," still occurring within clock time but not experienced that way subjectively.

Athletes call this "the zone." Writers call it "channeling." Musicians call it "losing yourself in the music." It's the same phenomenon, absorption so complete that conventional time awareness suspends.

Characteristic 4: Expanded Perception

You notice things you normally miss. Colors seem more vivid. Sounds more distinct. Details more apparent. Connections between things more obvious. This isn't hallucination, you're seeing what was always there but usually filtered out by your programming.

Remember: your brain filters out 99.9996% of sensory information before it reaches consciousness. During glitches, that filtering changes. Your brain shows you different aspects of reality, often things more relevant to your current needs or questions.

This explains why solutions to problems often appear during glitch states. The solution was always there in available information, you just couldn't see it through your normal filtering system. The glitch changes the filters, allowing you to perceive what was invisible before.

Characteristic 5: Effortless Performance

Actions that normally require intense effort feel natural and easy. Not because you're working less hard, but because you're no longer fighting your own programming while attempting the action.

Think about learning any skill. In the beginning, it requires intense conscious effort and feels awkward. Once mastered, it becomes effortless and automatic. Glitch states provide temporary access to that effortless performance even before you've fully mastered something, because they suspend the programming creating the awkwardness and difficulty.

This is why people in flow states often produce their best work, not because they're trying harder but because they're not interfering with their natural capabilities through excessive self-consciousness and programming interference.

Characteristic 6: Present-Moment Focus

During glitches, you're completely present in the current moment. Not thinking about past or future, not ruminating or planning, just fully engaged with what's happening right now.

This present-moment focus is both a cause and effect of glitch states. Being fully present helps trigger glitches by reducing prefrontal interference. Once in a glitch state, you naturally stay present because past and future concerns feel irrelevant.

Normal consciousness splits attention between experiencing the present moment and thinking about it, simultaneously doing something and having thoughts about doing it. During glitches, that split collapses. You're just doing, just experiencing, just being, without the extra layer of commentary.

Characteristic 7: Reduced Fear and Anxiety

Normal anxieties and fears significantly diminish or disappear entirely. Not because you're suppressing them but because they genuinely feel less relevant or threatening.

This happens because anxiety and fear are mostly generated by your programming, not by actual present-moment threats. They're predictions about future problems or reactions to past trauma. When you're completely present and your programming is temporarily suspended, those generated anxieties have nothing to hook onto.

This explains why people often perform courageously during glitch states. It's not that they're forcing themselves to be brave despite fear, the fear programming simply isn't running.

The Glitch Recognition Protocol

The single most important skill for exploiting glitch states is learning to recognize them as they occur. Most people experience glitches regularly but never consciously identify them, so they can't deliberately use them for transformation.

For the next two weeks, implement this recognition protocol. It will train your awareness to identify glitch states as they happen rather than only realizing afterward that something unusual occurred.

Hourly Check-Ins

Set a reminder on your phone to go off every two hours during waking hours. When it goes off, pause whatever you're doing and ask these questions:

Am I experiencing any unusual clarity or insight right now? Is my mental chatter quieter than normal? Does my sense of self feel different, more fluid, less bound by usual limitations? Has time been moving strangely, faster or slower than usual? Am I noticing things I usually miss? Do my usual anxieties feel less relevant or pressing? Am I completely absorbed in what I'm doing?

If you answer yes to three or more of these questions, you might be in or approaching a glitch state. This is the crucial moment: recognize it consciously. Actually think or say: "I'm in a glitch state right now. My normal limiting programming is suspended. I can use this."

The recognition itself helps stabilize and deepen the state. You're consciously acknowledging that something different is happening, which prevents you from unconsciously reverting to normal consciousness.

Daily Glitch Journal

Each evening, spend five minutes documenting any moments during the day that might have been glitch states, even if you're not certain. Write down:

What was happening when the potential glitch occurred? What did it feel like subjectively? What specific characteristics were present? How long did it last? What ended it? What did you accomplish or realize during that state? What could you have used it for if you'd recognized it earlier?

This daily practice trains your brain to recognize glitch state characteristics. After two weeks of journaling, you'll become significantly more sensitive to these states as they occur.

Pattern Analysis

After two weeks of hourly check-ins and daily journaling, review all your entries and look for patterns:

What activities or circumstances most commonly preceded glitch states? What time of day do they occur most frequently? What emotional or physical states correlate with them? How long do they typically last? What tends to end them?

This analysis reveals your personal glitch triggers, the specific conditions that increase likelihood of these states for you. Everyone's triggers are slightly different based on their neurology, psychology, and life circumstances.

Some people glitch most easily during intense physical activity. Others during creative work. Some during conversations with specific people. Others during solitary reflection. There's no universal trigger, which is why personal pattern analysis is essential.

Natural Glitch Triggers

While everyone's triggers are unique, certain activities and circumstances increase glitch state likelihood for most people. Understanding these helps you deliberately create conditions favorable to glitches.

Physical Intensity

Pushing your body significantly beyond its comfort zone often triggers glitch states. This happens because intense physical demand creates neurochemical changes and forces you into present-moment awareness while reducing prefrontal interference.

The runner's "second wind" is a glitch state. The clarity that comes during or after intense workouts is a glitch state. The expanded awareness that follows cold plunges or breath work is a glitch state. These aren't just endorphins making you feel good, they're windows where your normal programming temporarily suspends.

I experienced this dramatically during a challenging hike in the mountains. The first five miles were brutal, steep grade, high altitude, questioning why I'd thought this was a good idea. Around mile six, something shifted. The difficulty remained, but my relationship to it changed completely. I stopped fighting the challenge and started flowing with it. My usual mental chatter about discomfort and whether I could finish disappeared. I entered a state of calm, focused presence where each step felt perfect.

In that state, I had a series of insights about a professional challenge I'd been struggling with for months. The solutions appeared fully formed, obviously correct. I pulled out my phone and recorded voice notes because I knew from experience that glitch-state insights fade quickly if not captured.

When I returned home and listened to those recordings, the solutions still made perfect sense and proved successful when implemented. This wasn't just feel-good motivation from exercise, it was genuine problem-solving that happened because my normal limiting programming had temporarily suspended during the glitch state.

To use physical intensity as a glitch trigger, you need to push genuinely beyond comfort. Easy exercise won't do it. You need to create enough demand that your system shifts into a different mode of operation. This could be high-intensity interval training, long-distance running or cycling, challenging hikes, martial arts training, or any activity that pushes your limits.

The key is finding the sweet spot, challenging enough to trigger a state shift, but not so overwhelming that you can't maintain it. This is individual and requires experimentation.

Cold Exposure

Deliberate cold exposure, cold showers, ice baths, winter swimming, creates glitch states through powerful physiological responses. Cold triggers massive releases of norepinephrine and dopamine, creates intense present-moment focus by demanding complete attention, interrupts normal thought patterns through physical intensity, and activates your sympathetic nervous system in controlled ways.

The Wim Hof method explicitly uses cold exposure combined with breathing techniques to access altered states of consciousness. Athletes use ice baths for recovery but often report mental clarity and perspective shifts as valuable as physical benefits.

I started cold shower practice after reading about its benefits for resilience and mental clarity. The first week was genuinely difficult, fighting every instinct to avoid the discomfort. But around day eight, something changed. The cold became interesting rather than just unpleasant. I noticed I could watch my mind's resistance without being completely identified with it.

More importantly, I found that thirty seconds of cold water reliably shifted my state. If I was stuck on a problem or running old anxiety programming, a cold shower would reset everything. I'd emerge not just physically refreshed but mentally clear, with the mental chatter significantly quieted and often with new perspectives on whatever I'd been struggling with.

This isn't just about building physical toughness. The cold creates a controlled stress that suspends normal programming temporarily. In that suspension, you get a window to access different patterns of thinking and being.

Start gradually if you're new to cold exposure. End normal showers with thirty seconds of cold water, gradually increasing duration. Work up to full cold showers or ice baths only after building tolerance. The goal isn't suffering, it's creating conditions for state shifts.

Breathing Practices

Specific breathing patterns can trigger glitch states by altering blood chemistry, activating or calming your nervous system, creating altered consciousness states, and interrupting automatic thought patterns.

The Wim Hof breathing method involves cycles of deep breathing followed by breath retention. This creates significant changes in blood pH and oxygen saturation, often triggering altered states. Holotropic breathwork, developed by Stanislav Grof, uses specific breathing patterns to access non-ordinary consciousness states. Even simpler practices like box breathing or extended exhales can create noticeable state shifts.

I was initially skeptical about breathing practices creating genuine cognitive changes. It seemed too simple, just breathing differently couldn't really alter consciousness significantly, could it?

Then I experienced it directly during a Wim Hof workshop. After three rounds of the breathing protocol, forty deep breaths followed by extended breath retention, I entered a state unlike anything I'd experienced without external substances. My normal sense of boundaries became fluid. I felt simultaneously more aware and less self-conscious. Time became elastic. When the facilitator asked us to hold our breath after the third round, I held it for over three minutes without discomfort, something impossible for me in normal consciousness.

More significantly, I had insights during that altered state that proved valuable when I returned to normal consciousness. Problems I'd been overthinking suddenly seemed simple. Decisions I'd been agonizing over became obvious. Patterns in my behavior I'd been blind to became clearly visible.

Breathing practices work because breath directly affects your nervous system and brain chemistry. You're not just imagining changes, you're creating measurable physiological shifts that alter consciousness.

Extended Fasting

Strategic fasting can trigger glitch states through metabolic changes. When your body shifts from glucose metabolism to ketone metabolism during extended fasting, many people report increased mental clarity, reduced anxiety, altered time perception, and sometimes euphoria.

This happens because ketones affect brain chemistry differently than glucose. Your brain actually operates somewhat differently when running primarily on ketones, which can create subjectively different experiences of consciousness.

Religious traditions have used fasting to access altered states for millennia. While they described this in spiritual language, the mechanism is biological, changing your metabolic state changes your consciousness.

I practice intermittent fasting regularly, usually eating within an eight-hour window. But several times per year, I do extended fasts of twenty-four to seventy-two hours specifically to access the mental clarity and state shifts they create.

During these fasts, I notice increased cognitive sharpness around hour eighteen through twenty-four, emotional stability improvements and reduced reactivity, creative insights and problem-solving improvements, and what can only be described as spiritual or consciousness-expansion experiences.

These aren't just hunger-induced hallucinations. They're genuine state shifts that occur when your metabolism and therefore your brain chemistry operates differently than usual.

Important caveats: extended fasting should be approached carefully, especially if you have any health conditions. Consult medical professionals before attempting fasts beyond twenty-four hours. Start with intermittent fasting and build up gradually. The goal isn't suffering or deprivation, it's accessing altered states safely and sustainably.

Deep Meditation

Sustained meditation practice reliably triggers glitch states by quieting the default mode network, reducing self-referential thinking, creating gaps in automatic thought patterns, and accessing deeper levels of consciousness.

Long-term meditators report states that match glitch characteristics perfectly: quiet mind, altered time perception, identity fluidity, present-moment focus, expanded perception. These aren't mystical experiences, they're predictable results of specific practices that alter brain function in measurable ways.

Brain imaging of experienced meditators shows reduced default mode network activity, increased connectivity between brain regions, altered brain wave patterns, and changes in brain structure over time with consistent practice.

You don't need years of practice to access basic meditation-induced glitch states. Even twenty minutes of focused practice can create noticeable shifts if approached correctly.

The key is understanding that meditation isn't about stopping thoughts or achieving some special state. It's about observing your mind without identifying with or being controlled by its contents. This observation practice naturally creates gaps in normal programming's continuous execution.

I meditate daily for twenty to thirty minutes, using basic breath awareness or open monitoring practices. I don't do this primarily for the meditation session itself, I do it because consistent practice makes me more likely to access glitch states during the rest of my day.

The meditation works like exercise for your consciousness. Just as regular physical exercise makes your body more capable during daily activities, regular meditation makes your mind more capable of accessing useful states during daily challenges.

Intense Learning

Immersive learning experiences can trigger glitch states by overwhelming existing knowledge structures, creating cognitive dissonance, forcing new pattern recognition, and expanding capabilities rapidly.

Multi-day intensive workshops, complete immersion in new subjects, learning new languages intensively, and solving complex novel problems all create conditions favorable to glitches.

This happens because intense learning forces your brain to reorganize. When new information contradicts or expands beyond existing schemas, your brain has to restructure how it understands things. During this restructuring process, normal patterns become temporarily unstable, creating opportunities for glitch states.

I experienced this during a week-long medical training intensive where we learned a complex new surgical technique. The first two days were frustrating, the technique seemed impossibly complicated, my hands wouldn't cooperate, I couldn't integrate all the simultaneous steps.

Then on day three, something clicked. Suddenly I could see how all the pieces fit together. My hands knew what to do without conscious direction. The technique that seemed overwhelmingly complex suddenly seemed elegant and simple. I'd entered a glitch state created by the intensity of the learning process.

That state lasted for the rest of the intensive and carried forward. The technique remained accessible afterward because I'd learned it during a glitch state when my programming was more flexible and my brain more plastic.

To use learning as a glitch trigger, the key is intensity and novelty. Casual study won't create the necessary conditions. You need immersive, challenging engagement with genuinely new material that pushes your current understanding.

Creative Flow

Deep creative work can trigger glitches by demanding complete absorption, balancing challenge and skill, creating intrinsic motivation, and suspending self-consciousness.

Writers, artists, musicians, and creators across all domains report flow states that match glitch characteristics: time disappears, effort becomes effortless, self-consciousness vanishes, work emerges seemingly without conscious deliberation.

Mihaly Csikszentmihalyi's research on flow identifies conditions that trigger these states: clear goals with immediate feedback, balance between challenge level and skill level, complete focus on present-moment activity, intrinsic motivation rather than external rewards, and elimination of distractions.

These conditions don't guarantee flow, but they make it significantly more likely. The key insight is that you can create conditions favorable to flow rather than just waiting for inspiration to strike.

I experience creative flow most reliably during writing when I've warmed up with easier work first, eliminated all distractions completely, have clear goals but no attachment to specific outcomes, and am writing for inherent interest rather than obligation.

During flow, writing doesn't feel like effort. Words appear fully formed. Sentences structure themselves. Ideas connect without conscious deliberation. Time becomes irrelevant. I look up after what feels like thirty minutes to discover three hours have passed and I've produced work that would normally take days.

This isn't mystical, it's your brain operating without normal programming interference. The capability was always there; flow removes the obstacles that usually prevent accessing it.

Creating Glitch-Favorable Conditions

Understanding natural glitch triggers allows you to deliberately create conditions that increase glitch likelihood. This doesn't mean you can force glitches, they're spontaneous by nature. But you can significantly increase the probability of their occurrence.

The Preparation Phase

Glitches are more likely when you've created the right internal and external conditions. This preparation phase can be implemented anytime:

Clarify your intention. What do you want insight about? What problem needs solving? What capability do you want to access? Don't be desperate or attached, but have clear focus.

Eliminate distractions completely. Turn off phone notifications. Close unnecessary browser tabs. Remove anything that could pull your attention away. Glitches require complete present-moment focus.

Create physical comfort but not complete ease. You want to be comfortable enough to focus but challenged enough to stay alert. Slight discomfort or challenge helps trigger state shifts.

Set aside adequate time. Glitches can't be rushed. Block off at least two to three hours with no commitments afterward. The pressure of limited time prevents the letting-go that glitches require.

Release attachment to specific outcomes. Paradoxically, desperately wanting a glitch prevents it. Approach with interest and openness rather than grasping need.

The Trigger Phase

Once preparation is complete, engage in your chosen trigger activity:

For physical triggers: Push beyond comfortable intensity into genuine challenge. Find the edge where your body demands complete attention.

For cold exposure: Stay with the initial discomfort without fighting it. Watch your mind's resistance without being controlled by it.

For breathing: Follow the protocol consistently without trying to force anything. Let effects emerge naturally.

For meditation: Focus on present-moment awareness without getting frustrated when mind wanders. The practice is returning attention, not never losing it.

For learning: Immerse completely in the material. When confusion arises, stay with it curiously rather than frustrated.

For creative work: Start creating without judging what emerges. Lower the bar for initial output to reduce self-consciousness.

The key across all triggers is maintaining focus while releasing effort. You're creating conditions and then allowing whatever wants to emerge rather than forcing specific results.

The Recognition Phase

This is where most people fail. They experience glitches but don't consciously recognize them, so they can't deliberately use them.

As you engage in your trigger activity, maintain part of your awareness for recognizing when your state shifts. You're looking for the characteristics we identified earlier:

Is your mental chatter quieter than normal? Does time feel different? Do you feel more present and less self-conscious? Are insights emerging without deliberate thinking? Does your usual sense of limitation feel less real?

When you notice these shifts, consciously acknowledge them: "I'm entering a glitch state. My normal programming is suspending. I can use this for transformation."

Don't get excited about the glitch, excitement often ends it by reactivating normal programming. Stay curious and present instead.

The Exploitation Phase

Once you've recognized you're in a glitch state, deliberately use it for programming upgrades:

Install new beliefs that normally feel fake. In glitch states, new beliefs can install cleanly because your security software is offline. Visualize yourself already operating from upgraded programming. In glitch states, this visualization is more effective because you can access the experience more fully.

Make important decisions you've been stuck on. Glitch states provide clarity that normal consciousness lacks. Solve problems you've been struggling with. Solutions often appear obviously during glitches.

Practice capabilities you're developing. In glitch states, you can often access capabilities before you've fully mastered them, which accelerates learning.

The Integration Phase

This is the most critical phase. Glitches are temporary by nature, your normal programming will return. The question is whether you capture the insights and changes before normal consciousness resumes.

Immediately after a glitch state ends, within the first hour if possible, document everything. Write down insights, realizations, decisions, and new perspectives in detail. Your memory of glitch states fades quickly once normal consciousness returns.

Within twenty-four hours, take at least one concrete action based on your glitch-state insights. This tells your system that glitch-state intentions translate into real-world actions, making future glitches more impactful.

Over the following week, reinforce any programming you installed. Spend ten to fifteen minutes daily visualizing the new beliefs or capabilities as if they're already true. Take daily actions aligned with your upgraded programming. Notice and document evidence supporting your new beliefs.

This integration work determines whether your glitch becomes a one-time interesting experience or a genuine transformation. Without integration, insights fade and you return completely to old programming. With integration, glitches can create permanent upgrades.

Advanced Glitch Exploitation Techniques

Once you've experienced basic glitch recognition and exploitation, you can develop more sophisticated techniques that amplify their transformative power.

Glitch Stacking

This technique involves combining multiple glitch triggers simultaneously to create deeper, more stable altered states. The idea is that each trigger creates partial programming suspension, and combining them creates more complete suspension.

Thomas discovered this accidentally during a men's retreat in the mountains. The retreat combined intense morning exercise, cold plunges, breathwork sessions, extended periods of silence, and minimal food, all in a novel environment away from normal life. Each element alone might have created mild state shifts. Combined, they created profound glitch states that lasted for days.

During that retreat, Thomas experienced insights about his life and purpose that completely reoriented his trajectory. He'd been grinding in a corporate job he hated, telling himself he couldn't afford to change careers, carrying resentment toward his wife for not understanding his unhappiness, and generally feeling trapped by circumstances.

In the glitch states the retreat created, all of that programming temporarily dissolved. He saw clearly that his job dissatisfaction stemmed from doing work misaligned with his values, not from the work itself being inherently wrong. He recognized that his resentment toward his wife was projection, he was angry at himself for not having the courage to change, and blaming her was easier than taking responsibility. He understood that his sense of being trapped was programming, not reality, he had far more options than his scarcity mindset allowed him to see.

More importantly, in those glitch states, he could access a version of himself that already had the courage to change. Not as wishful thinking or aspiration, but as direct experience. For several days during the retreat, he simply was the person he wanted to become, confident, purposeful, aligned, free.

When he returned home, normal programming tried to reassert itself. His old identity whispered: "That was just the retreat high. Real life is different. You can't actually change." But Thomas had experienced too complete a glitch to fully believe those old patterns. He'd tasted what was possible when his limiting programming suspended entirely.

Over the following six months, he transitioned from corporate work to consulting in an area he genuinely cared about, rebuilt his relationship with his wife based on honesty rather than resentment, and created a life that felt authentic rather than obligatory. The retreat hadn't given him anything he didn't already have, it had shown him what was possible when his programming stopped interfering.

To practice glitch stacking, combine two or more triggers simultaneously:

Exercise intensely, then immediately do breathwork or cold exposure. Fast while engaging in intensive learning or creative work. Meditate in novel environments during life transitions. Combine physical challenge with creative practice. Layer multiple state-shifting practices in compressed timeframes.

The key is not overdoing it, you're creating conditions for glitch states, not trying to overwhelm yourself into exhaustion. Find combinations that feel challenging but sustainable.

Glitch Anchoring

This technique creates associations between physical cues and glitch states, allowing you to access altered states more reliably through conditioning.

The concept comes from neurolinguistic programming and classical conditioning: if you consistently pair a specific physical action with a specific mental state, eventually the physical action can trigger the state.

Jennifer developed this practice over several months. During spontaneous glitch states, she would touch her thumb and middle finger together in a specific way. Just a simple gesture, but one she only made during glitches. Over time, this gesture became associated with the glitch state in her nervous system.

After three months of consistent pairing, she discovered she could trigger mild glitch-like states by making the gesture deliberately. Not as profound as spontaneous glitches, but enough to shift her state noticeably, quieting mental chatter, increasing present-moment awareness, reducing anxiety.

She began using this anchor strategically before important meetings, challenging conversations, or creative work sessions. The gesture itself didn't create magic, it triggered her nervous system's memory of glitch states, which created partial access to those states.

To develop glitch anchoring:

Choose a specific, unusual physical gesture or posture you don't normally use. During glitch states, make this gesture consistently and hold it for at least thirty seconds. Pair the gesture with glitch states at least twenty times over several months. Begin testing whether the gesture can trigger mild glitch-like states. Use the anchor strategically in situations where you want to access altered states.

This won't give you on-demand access to deep glitch states, those remain somewhat spontaneous and resistant to forcing. But it can provide reliable access to mild state shifts that improve performance and perspective.

Glitch Journaling

This practice involves detailed documentation of every glitch experience to identify patterns and refine your personal glitch technology.

After experiencing my chemistry breakthrough glitch, I started keeping a detailed journal of every glitch state I experienced or suspected. For each entry, I documented:

What was happening when the glitch occurred? What triggers or conditions preceded it? What did the glitch feel like subjectively? What specific characteristics were present? How long did it last? What insights or capabilities emerged? What ended the glitch state? How did I feel immediately after? What did I do with the insights or changes?

After six months of detailed journaling, I had over forty documented glitch experiences. Analyzing these entries revealed patterns I never would have noticed otherwise:

My glitches most commonly occurred between 9 PM and midnight, during windows of high cognitive demand followed by relaxation. Physical exhaustion often preceded them, not random exhaustion but the specific tiredness that comes from pushing genuinely beyond normal limits. Novel environments increased glitch frequency significantly, traveling, new locations, changed routines. Conversations with specific people seemed to trigger glitches more than others. Certain types of problems or questions seemed to invite glitches, while others didn't.

This analysis let me deliberately create conditions similar to those that had spontaneously generated glitches. I couldn't force them, but I could significantly increase their frequency.

More importantly, the journaling practice itself seemed to increase glitch frequency. The act of paying attention to these states, documenting them, treating them as valuable rather than dismissing them, this attention itself appeared to make them more common.

Keep your glitch journal for at least three months before analyzing patterns. You need adequate data to identify genuine patterns versus random noise. After three months, review all entries looking for commonalities in triggers, timing, conditions, and characteristics.

Glitch Buddies

This advanced practice involves partnering with someone else who understands glitch states to provide mutual support, recognition, and accountability.

Marcus and David met at a workshop on consciousness and immediately recognized kindred spirits, both were interested in systematic personal development, both had experienced transformative glitch states, both wanted to learn to access them more deliberately.

They started meeting weekly for what they called "glitch sessions." Each session followed a loose structure:

They'd check in about glitch experiences from the past week, sharing details and insights. They'd discuss what they wanted to work on or access during the session. They'd engage in glitch-triggering activities together, breathwork, cold exposure, intense conversation, creative challenges. They'd share real-time observations about state shifts as they occurred. They'd document insights and commit to specific actions. They'd follow up the next week on whether they'd integrated previous insights.

Having a glitch buddy served multiple functions. It provided accountability, knowing he'd report to David motivated Marcus to actually implement insights rather than just having interesting experiences. It offered external perspective, David could often recognize when Marcus was in a glitch state before Marcus recognized it himself, helping him capture more glitches.

It created shared experiences, doing glitch-triggering activities together was both more effective and more sustainable than solo practice. It prevented spiritual bypassing, David would call Marcus out when he used glitch-state insights to avoid practical responsibilities rather than enhance them.

Most importantly, it normalized experiences that most people never discuss. Instead of feeling weird or delusional for paying attention to altered states, they had shared language and mutual understanding.

To find a glitch buddy, look for someone who demonstrates genuine interest in consciousness development, takes personal growth seriously without making it their entire identity, has their practical life reasonably together, can provide honest feedback without judgment, and shares your basic worldview enough to communicate effectively while different enough to provide valuable perspective.

This isn't about finding someone identical to you, it's about finding someone committed to similar exploration who can provide support, challenge, and accountability.

Glitch Documentation Templates

Having structured templates for documenting glitch experiences helps capture more useful information and identify patterns more easily.

I developed this template after months of unstructured journaling revealed I was missing important information:

Glitch State Documentation Template:
Date and Time: [When did this occur?]
Duration: [How long did the glitch state last?]

Location and Context: [Where were you? What were you doing?]

Preceding Conditions: [What happened in the hours before? Physical state? Emotional state? Activities?]

Trigger(s): [What seemed to initiate the state shift?]

State Characteristics: [Which glitch characteristics were present? Rate each 0-10]

Mental quiet: [0-10]

Identity fluidity: [0-10]

Time distortion: [0-10]

Expanded perception: [0-10]

Effortless performance: [0-10]

Present-moment focus: [0-10]

Reduced fear/anxiety: [0-10]

Primary Experience: [What was the most notable aspect of this glitch?]

Insights or Realizations: [What did you understand differently during this state?]

Capabilities Accessed: [What could you do that you normally can't, or what came more easily?]

Decisions Made: [What did you decide or commit to during the glitch?]

How It Ended: [What caused the state to shift back to normal consciousness?]

Immediate Integration Actions: [What did you do within 24 hours to integrate this experience?]

Longer-Term Impact: [Revisit this entry after one week and one month. What lasting impact did this glitch have?]

Using this template consistently creates a database of your glitch experiences that becomes increasingly valuable over time. After documenting twenty or thirty glitches using this structure, you can analyze the data scientifically rather than just having vague impressions about what works.

The Glitch States Paradox

Here's the paradox of glitch states that confuses most people: you can create conditions favorable to glitches, but you can't force them to occur. Trying too hard to trigger glitches actually prevents them. The optimal state is high intention combined with low attachment, you care deeply about accessing these states but don't desperately need any particular session to produce one.

This paradox explains why glitches often occur when you're not specifically trying to create them. You're engaged in some activity for its own sake, running because you enjoy running, creating because you love creating, learning because you're genuinely curious, and a glitch emerges spontaneously.

The moment you shift to "I'm doing this to trigger a glitch," you've introduced goal-oriented striving that often prevents the letting-go necessary for glitches.

Rachel struggled with this for months. She'd learned about glitch states and become obsessed with accessing them. She'd do breathwork desperately trying to force an altered state. She'd push through intense exercise while constantly monitoring: "Is this a glitch yet? How about now?" She'd sit in ice baths mentally commanding her consciousness to shift.

None of it worked. The harder she tried to force glitches, the more elusive they became. Her desperation itself created tension that prevented the state shifts she sought.

The breakthrough came when she temporarily gave up. She decided to just do the practices for their own sake without expecting specific results. She did breathwork because it felt interesting, cold exposure because it built resilience, exercise because it felt good. She released attachment to accessing glitches.

Within two weeks of this shift, glitches started occurring naturally. Once she stopped trying to force them, the conditions naturally emerged. She'd be running for the joy of running and slip into a glitch. She'd be writing for the love of writing and find herself in flow. She'd do breathwork just because and enter an altered state.

The key insight: glitches emerge from presence and letting-go, not from grasping and forcing. Create the conditions, engage fully with whatever you're doing for its own sake, and allow whatever emerges. This combination of high engagement and low attachment creates optimal conditions for glitches.

Glitch Ethics and Responsibility

As you develop capability with glitch states, you need to understand the ethical dimensions and responsibilities that come with this power.

Glitch states provide access to altered perspectives and capabilities. This power can be used for genuine growth and contribution, or it can be used to bypass responsibility, spiritual materialism (collecting experiences as achievements), or avoiding practical reality.

The Bypass Trap

Spiritual bypassing means using spiritual or consciousness practices to avoid dealing with practical responsibilities and psychological issues.

Kevin fell into this trap hard. He discovered glitch states and became obsessed with accessing them. He spent hours daily doing breathwork, meditation, cold exposure, anything that might trigger altered states. He had profound experiences, genuine insights, moments of expanded consciousness.

But his practical life was falling apart. He wasn't showing up for his job consistently. His relationships suffered because he was always "too deep in practice" to handle normal responsibilities. He used glitch-state insights to justify avoiding difficult conversations: "I realized during meditation that conflict is just ego, so I don't need to address problems with my partner."

This is bypassing, using consciousness exploration to avoid rather than enhance practical life. Real integration means glitch states make you more effective in normal reality, not less. They provide insights and capabilities you bring back to enhance your relationships, work, and responsibilities.

If your glitch practice is making you less functional in daily life, you're doing it wrong. The point isn't to escape reality, it's to upgrade your programming so you can engage with reality more effectively.

The Collection Trap

Spiritual materialism means treating consciousness experiences as achievements to collect rather than tools for genuine transformation.

Amanda became an experience collector. She'd done ayahuasca ceremonies, meditation retreats, breathwork intensives, float tanks, hypnosis, every consciousness-altering practice she could find. She loved talking about her experiences, comparing notes on how deep she'd gone, which practices produced the most intense states.

But nothing actually changed in her life. She had extraordinary experiences but never integrated them into practical transformation. She'd have profound insights during practices, then return to exactly the same patterns afterward. The experiences became ends in themselves rather than means to genuine growth.

Real glitch practice isn't about collecting interesting experiences. It's about using altered states to identify and upgrade limiting programming, then bringing those upgrades into daily life. The measure of a glitch's value isn't how profound it felt but how much it improved your actual functioning.

If you're having lots of glitch experiences but your practical life isn't improving, you're collecting rather than integrating. The solution is reducing experience-seeking and increasing integration work.

The Responsibility Framework

As you develop glitch capability, remember these principles:

Glitches are tools for upgrading your programming, not escapes from reality. The measure of success is practical improvement in your life, not the intensity of altered states. Integration work is more important than the glitches themselves. Your responsibility increases as your capability grows, use glitch-state insights to serve others and contribute, not just to benefit yourself.

Be honest about whether practices are helping you function better or helping you avoid functioning. Maintain balance between consciousness exploration and practical responsibility. Don't use glitch-state insights to justify avoiding difficult but necessary actions.

When Glitches Go Wrong

Sometimes glitch states can be destabilizing rather than helpful, especially if you're not prepared for them or don't have adequate integration support.

Intense glitch states can temporarily destabilize your sense of identity and reality. For people with certain psychological vulnerabilities, this can be genuinely problematic rather than just challenging. This is why traditional wisdom traditions always emphasized having a teacher or guide when exploring consciousness, and why modern psychedelic research includes extensive screening and professional support.

If you experience glitch states that feel destabilizing rather than expansive, that create persistent confusion rather than temporary uncertainty, that make it hard to function practically rather than easier, get professional support. Work with a therapist familiar with consciousness exploration and integration. Reduce or pause glitch-triggering practices until you've developed adequate grounding and support.

The goal is expanding your capabilities and consciousness while maintaining practical functionality, not destabilizing yourself. If your practices are creating problems rather than solving them, that's crucial feedback to adjust your approach.

The Glitch Mastery Journey

Developing genuine mastery with glitch states is a journey through predictable stages, each building on previous developments.

Stage 1: Unconscious Incompetence (Most People)

You experience glitch states but don't recognize them consciously. You have breakthrough moments, inspired insights, periods of unusual clarity, then return to normal consciousness without realizing something unusual occurred. You attribute these experiences to random luck, good moods, or mysterious inspiration.

Most people spend their entire lives in this stage, regularly accessing glitches but never learning to recognize or exploit them deliberately.

Stage 2: Conscious Incompetence (Weeks 1-4)

You've learned about glitch states and start recognizing them after the fact. You think: "Oh, that was a glitch state I was experiencing yesterday." You begin documenting experiences and looking for patterns, but you can't yet trigger them deliberately or exploit them while they're happening.

This stage is frustrating because you're aware of what's possible but can't consistently access it. The key is persistence, keep documenting, keep practicing, keep learning your personal patterns.

Stage 3: Conscious Competence (Months 2-6)

You can recognize glitch states as they occur and deliberately use them for programming upgrades. You understand your personal triggers and can create conditions favorable to glitches. You have protocols for exploiting glitches when they occur and integrating insights afterward.

This stage requires conscious effort, you're deliberately creating conditions, intentionally monitoring for state shifts, consciously implementing exploitation protocols. It works, but it's not yet automatic or effortless.

Stage 4: Unconscious Competence (Months 7-12)

Glitch recognition and exploitation becomes increasingly automatic. You naturally create conditions favorable to glitches without consciously thinking about it. You automatically recognize state shifts as they occur. You instinctively use glitch states for transformation without needing to follow conscious protocols.

This is where glitch work becomes integrated into your natural way of operating. You're not thinking "I should try to trigger a glitch," you're just living in ways that regularly create glitch opportunities, recognizing them naturally when they occur, and instinctively using them effectively.

Stage 5: Mastery (Year 2+)

Glitches become so frequent and integrated that the line between glitch states and normal consciousness begins to blur. You can access mild glitch-like states essentially on demand. Deep glitches still occur spontaneously but you can reliably create conditions for them.

More importantly, you've integrated enough glitch-state insights that your normal consciousness begins to operate more like glitch consciousness, quieter mind, present-moment focus, identity fluidity, expanded perception. You're not accessing special states as much as upgrading your baseline state.

This is the real goal: not to become dependent on accessing special states, but to upgrade your normal programming so thoroughly that your default consciousness approaches what used to only be accessible during glitches.

Your Glitch Development Plan

For the next four weeks, implement this systematic plan for developing glitch recognition and exploitation skills.

Week 1: Recognition Development

Focus entirely on learning to recognize glitch states without trying to trigger them deliberately. Set hourly reminders to check for glitch characteristics. Keep detailed daily glitch journal. Review this chapter daily to internalize glitch characteristics. Notice any experiences that might be glitches, even if you're uncertain.

Success metrics for week one: You can identify at least three potential glitch experiences. You understand the seven key characteristics. You've started recognizing patterns in when glitches occur. You're building awareness of state changes.

Week 2: Pattern Analysis

Continue recognition practice while beginning to analyze your personal patterns. Review week one journal entries for commonalities. Identify your most common glitch triggers. Notice times of day when glitches are most likely. Understand what conditions precede your glitches. Begin creating a personal glitch profile.

Success metrics for week two: You have clear hypotheses about your personal triggers. You know when you're most likely to glitch. You understand what conditions are favorable for you. You've identified at least two reliable patterns.

Week 3: Deliberate Triggering

Begin deliberately creating conditions similar to those that spontaneously generated glitches. Choose one trigger activity that seems most promising for you. Practice that trigger consistently this week. Maintain glitch recognition protocols while practicing. Document what happens during deliberate trigger attempts.

Success metrics for week three: You've practiced your chosen trigger activity at least five times. You've experienced at least one glitch during deliberate practice. You're understanding how to create favorable conditions. You're learning the balance between intention and attachment.

Week 4: Exploitation Practice

Focus on using any glitches that occur for deliberate programming upgrades. When glitches occur, consciously recognize and acknowledge them. Use glitch states to install specific belief upgrades. Make important decisions or solve problems during glitches. Document insights immediately after glitches end. Take concrete actions within 24 hours of glitch experiences.

Success metrics for week four: You've successfully exploited at least one glitch for programming upgrade. You've integrated at least one glitch-state insight into practical action. You're developing personal protocols for exploitation. You understand the complete cycle from recognition through integration.

After completing this four-week foundation, continue practicing and refining for months. Real mastery develops over years of consistent practice, but these four weeks establish fundamental skills that everything else builds on.

The Ultimate Glitch

Here's the final insight about glitch states that most people miss: your normal consciousness is actually the glitch. Your baseline state, with its mental chatter, identity constraints, limiting beliefs, and filtered perception, is the aberration, not the natural state.

What you experience during glitches, quiet mind, expanded awareness, present-moment focus, identity fluidity, effortless capability, that's closer to your actual natural consciousness before it was programmed with limitations.

Glitch states don't give you special powers or magic abilities. They temporarily remove the programming that's been blocking your natural capabilities. They show you what's been there all along, underneath the layers of conditioning, limiting beliefs, and automatic patterns.

This means the ultimate goal isn't learning to access glitches more frequently. It's upgrading your baseline programming so thoroughly that your normal consciousness begins to resemble what you currently only access during glitches.

When your usual state includes quiet mind, present awareness, identity fluidity, and expanded perception, you won't need to seek special altered states. You'll be living from an upgraded operating system where those capabilities are simply how you normally function.

This is what master-level reality hacking looks like: not someone who frequently accesses special states, but someone whose normal state has been upgraded to include what used to require glitches to access.

The glitches show you where you're going. They're previews of what becomes possible when you complete your programming upgrades. They're glimpses of who you actually are underneath the limiting code installed during your life.

Every glitch is the simulation showing you: "This is what's possible for you. This is who you actually are. This level of capability, clarity, and consciousness, this can be your normal state if you do the work to upgrade your programming."

The question isn't whether you can access glitches. You already do, regularly, whether you recognize them or not. The question is whether you'll recognize them, learn from them, and use them to systematically upgrade your reality operating system.

Your next glitch might happen today, tomorrow, next week. When it does, you'll be ready to recognize it, exploit it, and integrate it, and each glitch you successfully integrate brings you one step closer to making glitch-level consciousness your default state.

The simulation is glitching regularly, showing you what's possible. Are you paying attention?

PART 2: LEARNING TO HACK

Weeks 3-8: Exploiting Core Life Systems

Chapter 4: Money Code Exploits

Money is perhaps the most hackable system in your entire reality simulation. This might sound like an exaggeration until you understand what money actually is: a collective agreement, a shared belief system, a game with rules that can be learned and exploited. Unlike your body which has biological constraints, or relationships which involve other people's programming, money is pure code, and pure code can be pure hacked.

I discovered this during a moment of complete financial desperation. I was three months into my medical practice, barely covering expenses, wondering how I'd managed to get through years of training only to create a failing business. I had patients, I was providing good care, but money wasn't flowing the way I'd naively assumed it would once I had credentials and expertise.

Late one night, reviewing my dismal finances, I had a realization that changed everything: I wasn't struggling because I lacked skills or wasn't working hard enough. I was struggling because I was running poverty programming that my training had never addressed. Nursing school taught me science and clinical skills. It taught me nothing about money consciousness, value creation, or the psychological dimensions of financial success.

I was running code that said: "Money is hard to get and easy to lose." "Rich people are greedy or lucky." "I'm not supposed to care about money, I'm supposed to care about helping people." "Charging appropriately for my services is somehow wrong or greedy." "There's never enough money, no matter how hard I work."

This programming was generating my financial reality as surely as medical knowledge was generating my clinical outcomes. I could have the best clinical skills in the world, but if my money code was designed for scarcity and struggle, that's exactly what I'd experience.

The transformation began when I recognized my financial struggles as code execution rather than reality. My poverty wasn't a fixed external circumstance, it was the predictable output of programming I could identify and rewrite.

The Money Consciousness Hierarchy

Most people's money programming falls into one of several levels, forming a hierarchy from most limited to most abundant. Understanding where you currently operate is crucial for knowing what upgrades you need.

Level 1: Survival Consciousness

At this level, money feels scarce and threatening. You're constantly worried about having enough, afraid of losing what you have, and unable to think beyond immediate needs.

This was my starting point. Even after becoming a Nurse Practitioner and getting my PhD, I operated from survival consciousness around money. I had a scarcity mindset that made me hoard money fearfully, avoid spending even when investment would have yielded returns, underprice my services because charging appropriately felt dangerous, make decisions based on fear rather than strategic thinking, and treat every financial decision as potentially catastrophic.

Survival consciousness creates a self-fulfilling prophecy. When you operate from scarcity and fear, you make poor financial decisions. Poor decisions create poor outcomes. Poor outcomes reinforce your belief that money is scarce and threatening. The loop perpetuates itself.

The telltale signs of Survival Money Consciousness include constant anxiety about money regardless of how much you actually have, inability to spend money even on necessary investments or genuine needs, obsessive bargain-hunting even when time is more valuable than money saved, hiding or lying about finances due to shame, and automatically saying "I can't afford it" without considering whether you actually want it or how you might create resources for it.

Maria exemplified level one consciousness. Despite earning $65,000 annually, well above poverty level, she lived like she might become homeless any moment. She kept thousands in cash hidden in her apartment because banks felt unsafe. She'd drive thirty minutes to save three dollars on groceries. She wore clothes until they literally fell apart rather than buying replacements she could easily afford.

Her survival consciousness wasn't about her actual financial situation. It was programming installed during a childhood of genuine poverty and instability. That programming continued executing decades later in completely different circumstances, creating unnecessary suffering and limitation.

Level 2: Stability Consciousness

At this level, you're not constantly terrified, but you're focused primarily on maintaining what you have. You think in terms of security, safety nets, and avoiding risks.

Many middle-class professionals operate here. They've achieved some financial stability and their primary goal is protecting it. They save diligently, avoid risks, follow conventional wisdom, work hard in secure jobs, and measure success by accumulation rather than creation.

Stability consciousness is better than survival consciousness, at least you're not constantly panicked. But it's still fundamentally limited. You're playing defense rather than offense, protecting rather than building, avoiding loss rather than creating abundance.

The telltale signs include excessive focus on saving rather than investing or earning more, extreme risk aversion even for calculated opportunities, staying in limiting situations because they're "secure," following conventional financial advice without questioning whether it serves your goals, and measuring financial success by how much you've saved rather than what you've created.

Robert exemplified level two consciousness. He had a stable corporate job paying $95,000, maxed out his 401(k), had six months of expenses in savings, and felt proud of his responsible financial management. But he was deeply unhappy in his work, had entrepreneurial ideas he never pursued because they felt risky, and lived far below his potential because his stability consciousness couldn't tolerate the uncertainty of growth.

His consciousness wasn't wrong, stability is valuable, saving is important, security matters. But when stability becomes the primary goal rather than a foundation for creation, you sacrifice growth and possibility for the illusion of safety.

Level 3: Comfort Consciousness

At this level, you have enough money to live comfortably, and your primary focus is maintaining and gradually improving your lifestyle. You're no longer worried about survival or even stability, you're thinking about convenience, quality, and enjoyment.

Many successful professionals operate here. They earn good incomes, live well, can afford nice things, take vacations, and generally enjoy financial comfort. Their consciousness is focused on earning enough to support and slowly upgrade their lifestyle.

Comfort consciousness is pleasant but ultimately limiting. You're still trading time for money, even if the exchange rate is good. You're still thinking primarily about consumption rather than creation. You're still measuring success by what you can buy rather than what you can build.

The telltale signs include lifestyle inflation that matches income increases so you never build real wealth, focus on consumption rather than investment or creation, measuring success by possessions and experiences you can afford, working primarily to maintain lifestyle rather than build something meaningful, and difficulty imagining wealth significantly beyond your current level.

Jennifer exemplified level three consciousness. As a senior marketing director earning $150,000, she lived well, nice apartment, good restaurants, regular travel. She was comfortable and generally satisfied. But she worked long hours to maintain this lifestyle, spent most of what she earned, and couldn't imagine radically different possibilities because her consciousness was calibrated to comfortable employment.

Her consciousness wasn't wrong, enjoying life and treating yourself well is valuable. But when comfort becomes the ceiling rather than a baseline, you limit yourself to relatively modest success regardless of your capabilities.

Level 4: Success Consciousness

At this level, you're focused on achievement, growth, and building something significant. Money is important not primarily for what it buys but as a scorecard for value creation and a tool for building bigger things.

This is where serious wealth-building begins. Success consciousness focuses on creating value, building systems, leveraging resources, and playing bigger games. Money becomes abundant not because you're hoarding it but because you're consistently creating more value than you consume.

The shift from comfort to success consciousness is profound. Instead of asking "How can I earn enough to live well?" you start asking "What value can I create that others will pay for? How can I build systems that generate money while I sleep? What leveraged opportunities am I missing?"

The telltale signs include focus on value creation rather than just earning, building systems and assets rather than just working for income, thinking in terms of leverage and multiplication rather than linear exchange, measuring success by impact and creation rather than just personal consumption, and comfort with larger numbers and bigger possibilities.

Marcus exemplified level four consciousness. He transitioned from freelance designer earning $35,000 (comfort consciousness focused on surviving in his field) to business owner generating $240,000 (success consciousness focused on creating value and building systems).

The shift wasn't working harder, he actually worked fewer hours after the transition. It was changing his entire relationship with money from "scarce resource I earn through hours" to "abundant result of value creation through systems."

Level 5: Abundance Consciousness

This is the highest level, where money is simply energy that flows naturally as a result of contribution and creation. Scarcity thinking is completely absent. You operate from a fundamental knowing that resources are unlimited and you can create whatever you need.

At this level, money loses its emotional charge. It's important but not anxiety-producing. Abundant but not obsessed over. A tool for creation and contribution rather than a measure of worth or security.

Very few people operate here consistently, but those who do experience completely different financial realities. Money finds them. Opportunities appear. Resources materialize. Not through magic, but because their consciousness creates conditions where value flows freely and abundance is natural.

The telltale signs include complete absence of scarcity thinking or money anxiety, automatic focus on creation and contribution rather than acquisition, natural wealth generation through multiple channels, comfort with very large numbers and possibilities, and using money primarily as a tool for impact rather than personal consumption.

Dr. Sarah Chen exemplified level five consciousness. She built a thriving medical practice, launched a medical education company, invested in real estate and businesses, and served on nonprofit boards, all while maintaining health, relationships, and genuine enjoyment of life. When I asked about her financial success, she said something that revealed her consciousness: "I don't really think about money much. I focus on serving patients exceptionally, teaching effectively, and supporting causes I believe in. Money just flows naturally from doing those things well."

That's abundance consciousness, money as a natural result of contribution rather than something you struggle to acquire.

Identifying Your Money Code

Before you can hack your money programming, you need to see exactly what code is currently running. Most people have never examined their money beliefs consciously. They just run inherited programming and experience its outputs as "reality."

Take out a notebook and spend the next hour completing this money code audit. Be brutally honest. No one needs to see this but you, and self-deception here only limits your transformation.

Your Money Autobiography

Write down your earliest memories involving money. What did you observe about money in your family? What was said about money explicitly? What was communicated about money implicitly through actions and attitudes? What significant events shaped your money beliefs? What were you taught about wealthy people? What were you taught about poverty? What did you conclude about money and people like you?

This isn't therapy, it's debugging. You're tracing your current money code back to its installation points so you can understand why certain programs are running.

When I did this exercise, I remembered my mother crying over bills she couldn't pay, my father rage-spending money we didn't have then punishing everyone when it ran out, moving frequently because we couldn't afford rent, wearing clothes from charity organizations, and the constant underlying message that money was scarce, dangerous, and we were powerless to change our situation.

I could see clearly how those early experiences had installed code that said: "Money is scarce and hard to get." "Having money causes problems because someone will take it." "Asking for money is shameful." "People like us don't have money." This code was still running decades later, generating scarcity despite completely different circumstances.

Your Current Money Thoughts

For the next three days, carry a small notebook and write down every automatic thought you have about money. Don't try to change them, just observe and document.

You're looking for the automatic programming that runs below conscious awareness. These thoughts reveal the code generating your financial reality.

Common patterns include: "I can't afford that" (scarcity code). "Money doesn't grow on trees" (limitation code). "Rich people are greedy" (wealth-repelling code). "I'm not good with money" (capability limitation code). "There's never enough" (abundance-blocking code). "I don't deserve [expensive thing]" (worthiness code). "Money is the root of all evil" (wealth-demonizing code).

Write them all down without judgment. You're not trying to fix anything yet, you're just making your programming visible.

Your Money Decisions

Review the last ten significant financial decisions you made. For each one, write down: What was the decision? What were you trying to achieve or avoid? What fears or beliefs influenced the decision? What opportunities did you pursue or avoid? How did you feel before, during, and after the decision?

This analysis reveals your operating code through actions rather than just thoughts. What you actually do with money shows your real programming more accurately than what you say you believe.

When I did this analysis, I discovered that almost all my financial decisions were fear-based, avoiding risks, protecting what I had, staying safe. Even decisions that looked growth-oriented were actually fear-driven: I pursued certain opportunities not because they excited me but because I was afraid of missing out or falling behind.

Your Money Results

Look at your actual financial situation without judgment or justification. How much do you earn? How much do you save? How much debt do you carry? What's your net worth? How does money actually flow in your life?

Then ask the crucial question: What beliefs would I need to hold for these results to make sense?

If you earn $45,000 despite having skills worth $80,000, what belief would create that? Perhaps "I don't deserve more" or "Asking for more is greedy." If you spend everything you make regardless of income level, what belief creates that? Perhaps "Money disappears anyway so I might as well enjoy it" or "I'm not capable of managing money well."

Your results are outputs. Programming is the code generating those outputs. Working backward from results to code is one of the most powerful debugging techniques available.

Common Money Code Bugs

After working with hundreds of people on their money programming, I've identified recurring bugs that create predictable financial struggles. Recognizing these bugs in your own code is the first step toward fixing them.

Bug #1: The Scarcity Loop

This is the most common and destructive money bug. It runs continuously in the background, creating constant anxiety and poor decisions regardless of actual financial circumstances.

The code looks something like this: While money exists, worry about not having enough. Work harder to get more. Hoard what you have fearfully. Avoid spending even on genuine needs or investments. Feel temporary relief when you acquire more. Immediately restart worrying because it might run out. Repeat endlessly.

The bug creates an infinite loop where no amount of money ever feels like enough because the underlying code says money is fundamentally scarce. You could win the lottery and within months you'd be back to anxiety because your programming hasn't changed.

The telltale signs include constant money anxiety regardless of actual financial situation, inability to enjoy money you have because you're worried about future scarcity, automatically saying "I can't afford it" before considering whether you actually want something or how you might create resources, making decisions based on fear of loss rather than opportunity for gain, and feeling that having more money would solve all problems (it wouldn't, it would just make scarcity programming run on a larger scale).

The fix involves installing abundance recognition algorithms that notice unused resources and opportunities around you. Start documenting evidence of abundance daily, money that came to you, needs that were met, resources that appeared. Train your brain to filter for abundance as automatically as it currently filters for scarcity.

Bug #2: The Time-for-Money Trap

This bug limits income to available hours, creating the "busy but broke" syndrome where you work constantly but never build real wealth.

The code says: If you need money, sell hours. Work harder and longer to earn more. Value equals time invested. When you run out of time, you can't earn more. The only way to increase income is to work more hours or somehow charge more per hour.

This programming creates a hard ceiling on income. Even if you optimize hourly rates, there are only so many hours available. You're always trading time for money, which means you're always limited by time availability.

The telltale signs include measuring value by hours worked rather than results created, feeling guilty about earning money without proportional time investment, inability to imagine income beyond what you can personally produce through hours, resistance to systems or leverage that generate money without your direct involvement, and burnout from working more hours to try to earn more money.

The fix involves building systems that generate value independent of your direct time investment. This could be products, courses, content, investments, businesses with employees, or any model where value creation scales beyond your personal hours.

When I had this bug active, I could only imagine earning what I could produce through patient appointments. Fixing it meant developing group programs, online courses, and training systems that created value without requiring my direct time for every transaction. My income increased dramatically while my working hours decreased.

Bug #3: The Worthiness Gate

This bug prevents you from earning more than feels "appropriate" for someone from your background. It's like an invisible income ceiling that triggers guilt and self-sabotage when you approach it.

The code says: If income exceeds family historical level, trigger guilt. Question whether you deserve this success. Feel like you're betraying your roots or becoming someone you're not. Create problems or make poor decisions that bring you back down to familiar levels. Feel relief when you return to what feels "appropriate."

This programming ensures you never sustain income beyond what your identity can tolerate. You might temporarily earn more, but you'll unconsciously sabotage yourself back to your comfort zone.

The telltale signs include anxiety or guilt when earning significantly more than your parents or siblings, feeling like you're becoming a different person (negatively) when financial success increases, automatically creating problems or expenses that reduce your financial position when you're doing too well, discomfort with being seen as successful or wealthy, and difficulty sustaining income above a certain level despite having capability to earn more.

The fix involves redefining success as honoring rather than betraying your roots. Your success doesn't diminish others, it often creates opportunities for them. Earning well while maintaining your values proves it's possible to have both wealth and integrity.

I struggled with this bug intensely. Every time my income increased beyond what my parents had earned, I'd feel guilty and vaguely wrong. I'd make poor business decisions or create problems that brought me back down. The fix came from reframing: my success honored the sacrifices my parents made and proved their struggles weren't pointless. I was continuing their journey, not abandoning them.

Bug #4: The Money Virtue Code

This bug makes you believe that struggling with money somehow makes you a better person, while having money makes you morally suspect.

The code says: Poverty is noble, wealth is corrupt. Struggling with money means you're focused on what really matters rather than being greedy. Having financial success means you've compromised your values.

Real meaning comes from serving others regardless of compensation. Charging appropriately for your value is greedy or wrong.

This programming is especially common among helping professionals, creative people, and anyone raised with religious messages about money being the root of evil. It creates a no-win situation where you can be good or successful, but not both.

The telltale signs include pride in struggling financially as proof of your values, judgment toward people who charge premium prices or earn significant money, difficulty accepting money for work you enjoy or consider meaningful, feeling that money and meaning are incompatible, and unconscious patterns of giving away your value for free or grossly undercharging.

The fix involves recognizing that money is morally neutral, a tool that amplifies your existing character. If you're generous when poor, you'll be more generously impactful when wealthy. If you're greedy when poor, you'll be more destructively greedy when wealthy. Money reveals and amplifies what's already there, it doesn't corrupt good people.

I had this bug severely. As a medical professional, I'd internalized the message that caring about money meant not caring enough about patients. The breakthrough came from recognizing that charging appropriately allowed me to serve more sustainably and invest in better care, while undercharging led to burnout and reduced capacity to help anyone.

Bug #5: The Effort Equation

This bug makes you believe that money should be hard to get and that easy money is somehow wrong or invalid.

The code says: Money equals effort, the more you struggle, the more you should earn. Easy money is suspicious. If something pays well without being hard, it's probably wrong or won't last. Suffering and sacrifice should be rewarded financially. If you're not struggling, you're not really working.

This programming makes you unconsciously avoid opportunities that would make money easily or systemically. You stay stuck in models that require constant effort because easy abundance doesn't compute within your programming.

The telltale signs include guilt about earning money in ways that feel easy or enjoyable, belief that you should suffer for financial success, automatically making things harder than necessary, resistance to systems or models that generate money without constant effort, and valuing struggling employment over easier entrepreneurship despite lower returns.

The fix involves recognizing that value to others matters, not effort expended. The market pays for results and solutions, not for your struggle. If you can solve problems efficiently, that's more valuable than solving them laboriously.

I encountered this bug when I developed online training programs for medical professionals. They took significant upfront effort to create but then generated income repeatedly without my ongoing time investment. My programming said this was somehow wrong, I should have to work hard for every dollar. Fixing the bug meant recognizing that my value wasn't in time spent but in knowledge and systems I'd created.

The Money Code Upgrade Path

Once you've identified your money bugs and understood your current programming level, you can systematically upgrade toward abundance consciousness. This isn't about positive thinking or affirmations, it's about installing new code that generates different results.

Week 1: Scarcity Debug and Evidence Collection

Your first week focuses on debugging active scarcity programming and collecting evidence that contradicts it.

Every morning, spend five minutes visualizing abundance. Not as wishful thinking, but as systematic practice installing new neural patterns. Imagine money flowing to you easily. See yourself making financial decisions from confidence rather than fear. Feel what it would feel like to operate from abundance rather than scarcity.

Throughout the day, actively notice every instance of abundance: Unexpected income or gifts. Needs that get met without struggle. Resources that appear when you need them. Money-saving opportunities. Evidence that resources are more plentiful than your programming suggests.

Write these down in an abundance journal. You're training your brain to filter for abundance as automatically as it currently filters for scarcity. Your brain will show you whatever you consistently look for, currently it's calibrated to notice scarcity, but you can recalibrate it.

Also track your scarcity thoughts without judgment: Every time you think "I can't afford that" or "There's not enough" or any scarcity-based thought, write it down. You're not trying to stop these thoughts, you're just making the programming visible so you can eventually modify it.

By week's end, you'll have pages of documented abundance evidence showing that resources are more plentiful than your programming acknowledges, plus documentation of your active scarcity code.

Week 2: Worthiness Installation

Week two focuses on installing programming that says you deserve financial success and prosperity.

Create a worthiness statement that feels true to you. Not something that sounds good but triggers disbelief, but something that honestly reflects your growing understanding. Examples: "I deserve to be fairly compensated for the value I provide." "I am worthy of financial abundance and success." "Money flows to me easily because I create genuine value for others."

Spend ten minutes each morning repeating your worthiness statement while visualizing yourself operating from that belief. See yourself receiving money confidently, charging appropriate rates without guilt, making financial decisions from worthiness rather than unworthiness.

Throughout the day, collect evidence that supports your worthiness: Times you created value for others. Positive feedback on your work. Problems you solved. Skills you have. Capabilities you've developed. Ways you've helped people.

Write these down. You're building a counter-narrative to unworthiness programming by documenting objective evidence of your value.

Also practice micro-worthiness decisions: When facing small financial choices, deliberately choose based on worthiness rather than unworthiness. Buy the slightly nicer thing instead of always defaulting to cheapest. Invest in something meaningful rather than automatically saying you can't afford it. These small decisions train your system to operate from new programming.

Week 3: Value Recognition and Communication

Week three focuses on recognizing the full value you create and learning to communicate it effectively.

Make a comprehensive list of the value you provide: What problems do you solve? What needs do you meet? What results do you create? What transformations do you facilitate? What would people lose if you weren't available?

Be specific and honest. Most people dramatically underestimate the value they provide because they take their abilities for granted. Ask others what value you create, their perspective often reveals things you don't see.

Then practice communicating this value clearly: Write down how you would describe what you do in terms of value rather than tasks or time. Practice explaining your worth without apologizing or minimizing. Role-play conversations about money where you confidently express your value.

This might feel awkward or arrogant initially, that's your old programming resisting. Clearly stating your value isn't arrogance; it's accurate representation that serves everyone by creating appropriate expectations and exchanges.

Also begin the practice of thinking in terms of value rather than cost: When you see prices, consider the value rather than just the expense. "Is this worth it?" rather than just "Can I afford it?" This trains your brain toward abundance thinking focused on value exchange rather than scarcity thinking focused on resource depletion.

Week 4: Opportunity Recognition

Week four focuses on training your brain to notice and act on financial opportunities that your old programming filtered out.

Set an intention each morning: "I notice and act on opportunities today." This primes your pattern recognition for possibilities.

Throughout the day, actively look for opportunities: Ways to create value for others. Problems you could solve profitably. Unmet needs in your network or market. Resources you could leverage. Possibilities you normally dismiss without consideration.

When you notice opportunities, don't immediately dismiss them. Your old programming will say "That won't work" or "I can't do that" or "That's for other people." Instead, pause and consider: "What would it take to make this work? How might this be possible?"

Write down at least three opportunities daily, even small ones. Then choose at least one per week to act on in some way, research it, discuss it with someone, take a small step toward exploration.

You're training your brain to see possibility where it currently sees limitation, and training yourself to act on opportunities rather than automatically dismissing them.

Advanced Money Hacking Techniques

Once you've completed the basic four-week upgrade path, you can implement more sophisticated techniques that dramatically accelerate money code transformation.

Income Expansion Protocol

This technique systematically expands your comfort with larger numbers and higher income levels.

Your brain has a sort of "money thermostat" set to a particular income range that feels normal. When you exceed it, you unconsciously create problems or expenses that bring you back down. When you fall below it, you unconsciously create opportunities that bring you back up. The goal is to reset this thermostat to a higher level.

Start by choosing a target annual income that's beyond your current level but not so far beyond that it feels completely impossible. Perhaps 50-100% more than you currently earn. Write this number down and spend five minutes daily visualizing earning this amount.

Not as fantasy, but as systematic rehearsal: See yourself receiving payments at this level. Imagine your bank account showing these numbers. Feel what it would feel like to earn this much. Make it as real and detailed as possible. Experience the emotions, the confidence, the ease of operating at this level.

Then, crucially, start making decisions as if you already earned this amount: How would you allocate your time differently? What opportunities would you pursue? What would you say yes to, and what would you say no to? How would you price your services? What investments would you make?

Begin making actual decisions based on this future income level, within reason. You're training your system to operate from a higher financial baseline, which creates conditions for that baseline to materialize.

Marcus used this technique to shift from $35,000 to $240,000 in eighteen months. He visualized earning $150,000 (his initial target), then made business decisions as if he already earned that much: pricing services for $150,000 earner value, investing in business development, declining low-value projects, positioning himself as premium. His income rose to meet his programming.

Value Multiplication Exercise

This practice trains you to think in terms of leverage and multiplication rather than linear exchange.

Most people think: "If I work X hours at Y rate, I earn Z money." This is linear, limited thinking. Abundance thinking recognizes that value can be created once and sold many times, that systems can generate without your direct involvement, that leverage multiplies individual effort exponentially.

Each week, identify one area where you could multiply value: Could you create something once that serves many people repeatedly? Could you build a system that works without your constant involvement? Could you leverage other people's time, money, or expertise? Could you solve a problem for many people simultaneously rather than one-by-one?

Then take at least one concrete step toward implementing value multiplication. You don't need to build entire businesses, start small. Write one article that can be read thousands of times. Create one template others can use. Build one small system that reduces repetitive work.

The practice trains your brain to think in terms of multiplication and leverage rather than just trading time for money. Over time, this thinking leads to dramatically different financial results.

Your Money Transformation Timeline

Here's what successful money code upgrade looks like over twelve weeks:

Weeks 1-4: Foundation (Scarcity Debug)

You're seeing your money programming as code rather than reality for the first time. You're collecting abundance evidence that contradicts scarcity programming. You're experiencing moments where scarcity thinking automatically arises, but you can recognize it as programming rather than truth. You're beginning to make different small decisions based on emerging abundance consciousness.

By week four, you should notice reduced money anxiety, increased awareness of opportunities, small improvements in financial circumstances, and most importantly, the recognition that your money struggles are programming rather than fixed reality.

Weeks 5-8: Development (Value Embodiment)
You're installing worthiness and value-recognition programming. You're speaking about your value more confidently. You're charging more appropriately for your services. You're pursuing opportunities you would have dismissed before. You're making decisions from abundance rather than scarcity more consistently.

By week eight, you should see measurable income improvements, perhaps 15-30% increases from existing income sources, new income streams beginning to develop, reduced financial stress despite same or better financial position, and significantly improved relationship with money conversations.

Weeks 9-12: Integration (Abundance Stabilization)
Your new money programming is becoming automatic. Abundance thinking is your default more often than scarcity thinking. You naturally recognize and act on opportunities. You comfortably charge appropriate rates and communicate your value. Financial decisions come from strategy rather than fear.

By week twelve, you should have achieved 30-50% income increases through various channels, developed at least one new income stream, eliminated or dramatically reduced money anxiety, and established sustainable abundance consciousness that generates ongoing results.

These numbers aren't guarantees, individual results vary based on starting point, consistent implementation, and life circumstances. But they represent typical outcomes when people systematically upgrade their money programming over three months.

The key is consistency. Daily practice matters more than occasional inspiration. Five minutes every morning installing new programming creates more transformation than reading books occasionally or thinking about money when problems arise.

Your money code is always running, always executing, always generating your financial reality. The question isn't whether you have money programming, you do, and it's creating your current results. The question is whether you'll take conscious control of that programming and upgrade it to generate abundance rather than scarcity.

The simulation responds to whatever money code you're running. Change the code, change the results. It's really that simple, and that challenging. Simple because the principle is straightforward. Challenging because it requires confronting programming installed over decades and consistently practicing new patterns until they become automatic.

But it's possible. I've done it. Hundreds of people I've worked with have done it, and you can do it too, starting this week.

Your financial transformation doesn't require luck, special advantages, or magic. It requires recognizing your money programming, systematically debugging limitations, and installing code designed for abundance rather than scarcity.

The money is already there. The opportunities already exist. The resources are available. Your current programming just filters most of it out. Upgrade the programming, and you'll suddenly see what was invisible before.

Ready to rewrite your money code.

Chapter 5: Relationship Algorithm Hacks

Your relationships aren't random attractions or lucky connections, they're the predictable output of sophisticated algorithms running in your consciousness that determine who you attract, how you interact, and what patterns you create with others. These algorithms are so powerful that they don't just affect your happiness; they actively shape your reality by influencing your opportunities, resources, and what becomes possible in your life.

I discovered this the hard way during my second year of nursing school when I found myself in yet another relationship that was repeating the exact same patterns I'd experienced in all my previous relationships. Different person, identical dysfunction. I was attracted to emotionally unavailable men, created push-pull dynamics where I pursued and they withdrew, felt chronic anxiety about being abandoned, and eventually either got abandoned or left first to avoid the pain of being left.

This wasn't bad luck. This wasn't just picking the wrong people. This was code, sophisticated programming installed during childhood that automatically generated these patterns regardless of who I was dating.

The realization hit me during a particularly painful breakup. I was crying to a friend about how "all men" were emotionally unavailable, when she said something that changed my trajectory: "Pauline, you've dated multiple men in the past five years. They weren't all emotionally unavailable, you just couldn't feel attracted to the ones who were available. Your system filters for unavailability because that's what feels familiar."

She was right. I'd had opportunities to date men who were genuinely available, interested, and emotionally mature. But I couldn't feel that "spark" with them. The spark I felt, the intense chemistry and attraction, only activated with men who were distant, ambiguous, or already involved with someone else.

My relationship programming was running automatically below conscious awareness: scan for emotionally unavailable men because that's what feels like love, create anxiety and intensity because that feels like passion, pursue when they withdraw because that feels like devotion, and interpret their unavailability as a challenge to prove I'm worthy of love.

This code had been installed during childhood watching my mother chase my drug addicted father's attention, learning that love meant pursuing someone who couldn't fully show up, that anxiety meant you cared, that intensity meant it was real. That programming was still executing perfectly decades later, generating the exact same relationship dynamics my six-year-old self had learned.

Understanding this changed everything. My relationship struggles weren't personality flaws or communication problems. They were code execution. Until I debugged and upgraded my relationship algorithms, I would keep generating the same patterns regardless of who I dated or how hard I tried to make relationships work.

The Relationship Operating System

Just as you have a Reality OS governing your overall experience, you have a Relationship OS, a set of core programs determining how you relate to others. Most people are running relationship code installed during childhood based on their first experiences of attachment, love, and connection.

Attachment Programming: The Foundation Code

Your attachment style is the deepest layer of relationship programming, usually installed before age three based on how your primary caregivers responded to your needs. This foundational code determines how you approach all intimate relationships throughout your life.

Secure attachment code was installed when caregivers were consistently responsive, creating programming that says: relationships are safe and reliable, people are trustworthy, I'm worthy of love and attention, intimacy is comfortable and natural, I can depend on others while maintaining independence.

People running secure attachment code make relationships look easy. They're comfortable with both intimacy and independence, can navigate conflicts constructively, don't catastrophize during disagreements, and recover quickly from relationship problems. About 50% of people have this programming, which means 50% don't.

Anxious attachment code was installed when caregivers were inconsistently responsive, sometimes available, sometimes not, creating programming that says: relationships are unreliable and anxiety-producing, I need to pursue and cling to maintain connection, I'm not inherently worthy so I must earn love through effort, any distance means abandonment, I need constant reassurance and closeness.

People running anxious code experience relationships as chronically stressful. They need frequent reassurance, fear abandonment, interpret normal distance as rejection, pursue when partners withdraw, and feel they love more than they're loved. About 20% of people have this programming.

Avoidant attachment code was installed when caregivers were consistently unresponsive or intrusive, creating programming that says: relationships threaten my autonomy, depending on others is unsafe, I must be self-sufficient, too much closeness feels suffocating, emotional distance protects me.

People running avoidant code value independence highly, feel uncomfortable with too much intimacy, withdraw when partners seek closeness, rationalize emotional distance as maturity, and struggle to access or express vulnerable emotions. About 25% of people have this programming.

Disorganized attachment code was installed when caregivers were frightening or chaotic, sometimes nurturing, sometimes threatening, creating programming that says: relationships are both desperately needed and terrifying, I want closeness but fear it simultaneously, intimacy feels dangerous but so does isolation, my needs are shameful and wrong, I'm fundamentally unlovable.

People running disorganized code experience extreme relationship chaos, craving intimacy then sabotaging it, intense connections followed by explosive endings, difficulty trusting even people who prove trustworthy, and chronic confusion about what they want. About 5% of people have this severe programming, though many more have elements of it.

I was running primarily anxious attachment code with some disorganized elements, which explained my pattern of pursuing unavailable partners while simultaneously sabotaging available ones. The unavailable partners activated my anxiety programming perfectly, I could pursue endlessly without having to face the terror of actual sustained intimacy that might trigger my disorganized code.

Identifying Your Relationship Code

Before you can hack your relationship algorithms, you need to see exactly what code is currently running. Most people have never consciously examined their relationship patterns, they just keep repeating them and wondering why they can't find the right person.

Take out a notebook and complete this relationship code audit over the next week. This requires brutal honesty and potentially uncomfortable self-reflection, but accurate diagnosis is essential for effective debugging.

Your Relationship Pattern History

List your last five significant relationships or connections. For each one, document: What initially attracted you to this person? What patterns emerged in the relationship? How did conflicts typically unfold? What role did you play? What role did they play? How did it end? What similarities exist across all five?

Most people discover shocking consistency when they do this exercise. The specific people were different, but the patterns, conflicts, and dynamics were nearly identical. This reveals code execution rather than bad luck.

When I did this analysis, I discovered that every relationship followed the same script: intense initial attraction with someone emotionally unavailable, pursuit and anxiety when they withdrew, brief periods of closeness followed by distance, my increasing anxiety and their increasing withdrawal, eventual abandonment or my preemptive exit. The names changed, but the code executed identically every time.

Your Attraction Algorithm

Answer these questions honestly: What creates that "spark" or strong attraction for you? What qualities in others make you feel most drawn to them? What patterns do you notice in who you're attracted to versus who's attracted to you? Are you more drawn to people who are available or unavailable? Do you feel stronger attraction to people who pursue you or people you pursue? What does "chemistry" feel like in your body, and when does it occur?

The crucial insight: your "chemistry" or "spark" isn't random attraction, it's your code recognizing patterns that match your programming. If your programming says "love means pursuing someone distant," you'll feel intense chemistry with distant people and boredom with available ones.

I realized my strongest "chemistry" occurred with men who were ambiguous, inconsistent, and emotionally guarded, exactly the patterns my childhood programming recognized as love. Men who were straightforward, consistent, and emotionally available felt boring or friend-like. My attraction algorithm was literally filtering for the exact pattern that would recreate my childhood dynamics.

Your Conflict Patterns

How do you typically respond when conflicts arise in relationships? Do you pursue or withdraw? Attack or defend? Explode or shut down? Need to resolve immediately or need space? Apologize excessively or struggle to apologize? What triggers your strongest reactions? What do you absolutely need from others during conflicts? What's intolerable for you?

These patterns reveal your conflict code, the programs that execute automatically during relationship stress. Understanding them helps you recognize when code is running versus when you're responding consciously.

My pattern was pursuit during conflict, I needed immediate resolution, couldn't tolerate distance or silence, and would pursue relentlessly when partners withdrew. This behavior, driven by my anxious attachment code, actually created more distance by making partners feel overwhelmed and claustrophobic. My code was generating the exact outcome it was designed to prevent.

Your Boundary System

Do you have clear boundaries that you maintain consistently? Do you say yes when you want to say no? Do you accommodate others at your own expense? Do you feel guilty setting limits? Do you get resentful when people don't respect boundaries you never clearly set? Can you receive help and support or do you insist on handling everything alone?

Boundary patterns reveal whether you're running people-pleasing code, over-independence code, or healthy interdependence code. Most people with attachment issues have boundary problems in one direction or the other, either no boundaries (anxious) or walls instead of boundaries (avoidant).

I had virtually no boundaries with romantic partners. I would accommodate endlessly, hoping this would earn the love and security my code craved. Then I'd become resentful when my unstated needs weren't met, which would trigger arguments where I'd explode about things I'd never actually communicated bothering me. My lack of boundaries came from programming that said: "My needs aren't important, and stating them will drive people away."

Common Relationship Code Bugs

After years of working on my own relationship programming and helping others debug theirs, I've identified recurring bugs that create predictable relationship problems.

Bug #1: The Unavailable Attraction Algorithm

This bug causes you to feel intense attraction to emotionally unavailable people while feeling bored or "friend-zoned" by available, healthy people.

The code runs like this: Scan environment for emotionally unavailable people. Feel intense "chemistry" when you identify unavailability. Interpret their distance as mysterious and attractive. Feel bored or friend-like with emotionally available people. Convince yourself unavailable people are "worth the effort."

This bug typically installed during childhood when a caregiver was present but emotionally distant. Your young mind learned to associate love with pursuing someone who can't fully reciprocate. The anxiety of pursuit felt like passion. The challenge of earning unavailable love felt like proof of your worth.

Lisa exemplified this bug perfectly. A successful attorney, intelligent and self-aware, she kept falling for men who were "busy," "not ready for a relationship," "recently divorced and still processing," or otherwise clearly unavailable. She'd pursue them intensely, convinced that if she was patient and understanding enough, they'd eventually become available for real relationship.

Meanwhile, men who were actually available, interested, and emotionally mature felt "too nice" or "too eager." She couldn't feel that exciting chemistry with them. "I know I should date the available guys," she'd say, "but there's just no spark."

Her code was working perfectly, generating attraction to familiar patterns (unavailability) while filtering out unfamiliar patterns (availability). The "spark" she felt wasn't chemistry, it was recognition. Her system was identifying someone who matched her programming's definition of love.

The fix required debugging her equation of anxiety with passion and installing new code that recognized: True chemistry can exist with availability. Anxiety isn't love, it's anxiety. A "spark" with unavailable people is my code recognizing familiar dysfunction. Attraction to available people can develop when I don't immediately reject it. Comfort and security can coexist with passion and excitement.

She practiced deliberately dating available men even when the initial "spark" wasn't there, staying long enough to see if attraction could develop. After three months of this practice, she reported something remarkable: she'd started feeling genuine attraction to a man who was straightforward, consistent, and emotionally available. The attraction felt different than her usual intense anxiety, calmer, more grounded, but also more genuinely excited about him as a person rather than excited about the chase.

Bug #2: The Rescuer-Victim Protocol

This bug compels you to relationships where you're either rescuing someone or being rescued, rather than meeting as equals.

The code says: Scan for people who need help or fixing. Feel important and valuable when you're needed. Equate being needed with being loved. Feel uncomfortable with equal partnerships because they don't activate your value programming. Maintain rescuer role to maintain relationship.

Or alternatively: Scan for strong people who can rescue you. Feel safest when someone else is handling things. Equate being taken care of with being loved. Feel uncomfortable taking responsibility because dependency feels like love. Maintain victim role to maintain relationship.

This bug typically installed when childhood circumstances required you to be overly responsible (parentified child who becomes rescuer) or when you were treated as incapable and dependent (creating victim programming).

Marcus ran rescuer code intensely. Every woman he dated had significant problems, addiction, financial disasters, dramatic family situations, or emotional instability. He'd pour himself into helping them, convinced his love and support could fix them.

The relationships would follow a predictable pattern: He'd meet someone struggling. He'd feel drawn to "help." He'd invest enormous energy into their problems. They'd improve temporarily, sometimes significantly. Then either they'd leave once they felt stronger, or they'd resent his "helping" and rebel, or new problems would emerge to maintain the dynamic.

The pattern repeated because Marcus's code generated it. His programming said: "I'm valuable when I'm needed. If someone doesn't need rescuing, they won't need me. Therefore, I must find people who need rescuing to feel loved and important."

The fix required debugging his equation of need with love and installing new code: I'm valuable as a person, not just as a helper. Equal partnerships are more sustainable than rescue dynamics. People who need rescuing usually aren't ready for healthy relationship. Being wanted is healthier than being needed. My role is partner, not savior.

Marcus began deliberately dating women who didn't need rescuing, who had their lives together, who could handle their own problems, who wanted a partner rather than a rescuer. Initially this felt wrong; his code said these relationships couldn't be real because he wasn't needed. But as his new programming installed, he discovered something profound: being wanted as an equal partner created deeper, more satisfying connection than being needed as a rescuer ever had.

Bug #3: The Rejection Sensitivity Algorithm

This bug interprets neutral or slightly negative responses as major rejections, leading to defensive behaviors that actually create the rejection you're trying to avoid.

The code runs: Scan constantly for signs of rejection or disapproval. Interpret ambiguous signals as negative. Feel threatened by normal distance or independence. React defensively, withdraw before you're rejected, attack when you feel criticized, or desperately pursue to prevent abandonment. Create the rejection you feared through your defensive response.

This bug creates self-fulfilling prophecies. Your fear of rejection triggers behaviors that push people away, confirming your belief that you'll be rejected.

Jennifer struggled with this bug intensely. Any sign that her boyfriend might be pulling away, a delayed text response, a quieter-than-usual mood, wanting a night alone, would trigger her rejection programming. She'd become anxious, needy, and accusatory, demanding to know what was wrong and insisting they talk immediately.

Her boyfriend would then actually pull away because her behavior felt overwhelming and intrusive. She'd interpret his withdrawal as confirmation of her worst fears, triggering even more desperate pursuit. The relationship eventually ended with him saying he felt "suffocated" and needed space, the exact rejection her programming feared and inadvertently created.

The fix required installing new interpretation code: Normal distance doesn't mean rejection. People need space sometimes without it meaning anything about the relationship. Ambiguous signals are probably neutral, not negative. My anxiety is code running, not accurate information. Giving space when I want to pursue creates better outcomes than desperate pursuit.

Jennifer learned to sit with her rejection anxiety without acting on it. When her programming said "He's pulling away, pursue immediately," she'd recognize it as code, feel the anxiety without obeying it, and create space instead of pursuing. Paradoxically, this reduced the distance she feared because her partners no longer felt overwhelmed by her pursuit.

Bug #4: The Intimacy Avoidance Subroutine

This bug creates distance whenever relationships become too close or vulnerable, often through picking fights, finding flaws, or creating crises that justify withdrawal.

The code says: If intimacy exceeds safe threshold, trigger distance-creation response. Find flaws in partner suddenly. Feel suffocated or trapped. Create conflict that justifies withdrawal. Feel relief when distance is re-established. Repeat when things get close again.

This bug typically installed when closeness wasn't safe in childhood, either from intrusive caregivers who violated boundaries, or from caregivers who used closeness to manipulate, or from traumatic experiences that made vulnerability dangerous.

David exemplified this bug. He'd pursue women intensely, charm them, create strong initial connection. Then once the relationship deepened and became more intimate, he'd start withdrawing. He'd suddenly notice flaws that bothered him. He'd feel "trapped." He'd create arguments or distance. Eventually he'd end things, feeling a sense of relief and freedom.

Then he'd repeat the pattern with someone new. He genuinely wanted relationship in theory, but couldn't tolerate actual sustained intimacy in practice. His code was protecting him from the vulnerability he unconsciously associated with danger.

The fix required understanding that his withdrawal wasn't about the relationships or the partners, it was defensive programming activating when intimacy triggered his safety code. He needed to install new programming: Intimacy is safe with trustworthy people. Vulnerability is strength, not weakness. Closeness doesn't mean losing myself. Flaws that suddenly appear are probably my code activating, not real dealbreakers. Staying present through discomfort builds genuine intimacy.

David learned to recognize when his avoidance programming activated, the sudden critical thoughts, the trapped feeling, the urge to withdraw, and stay present instead of automatically acting on it. This was profoundly uncomfortable initially; his entire system wanted to run the old program. But as he practiced staying with intimacy past his usual threshold, his code began to update. He discovered that closeness didn't actually destroy him the way his programming predicted.

Bug #5: The People-Pleasing Protocol

This bug prioritizes others' comfort over your own authenticity, creating relationships based on performance rather than genuine connection.

The code runs: Automatically sense what others want or need. Adapt yourself to provide it. Suppress your own needs, preferences, and boundaries. Fear disappointing or upsetting others. Measure your worth by how much others approve of you. Exhaust yourself maintaining the performance.

This bug installed when childhood approval was conditional on being who others needed you to be rather than who you actually were. You learned that authentic self was unacceptable, so performance self kept you safe.

Rachel ran this code perfectly. In relationships, she became whoever her partner wanted. Dating someone who valued fitness? She became obsessed with working out. Dating someone intellectual? She became fascinated by academia. Dating someone who wanted independence? She suppressed her needs for connection.

The problem: she had no idea who she actually was anymore. She'd spent so long performing that her authentic self was buried under layers of adaptation. Her relationships felt hollow because no one actually knew her, they knew her performance.

The fix required debugging her equation of approval with love and installing new code: I am lovable as myself, not just as a performance. Authenticity creates genuine connection. People-pleasing prevents real intimacy. My needs and preferences matter. Disappointing others occasionally is normal and healthy.

Rachel began the terrifying practice of expressing her authentic preferences, needs, and opinions even when they might disappoint or conflict with others. Her current relationship couldn't survive this shift, it was based entirely on her performance of who he wanted, but she eventually attracted a partner who valued her authenticity over her accommodation.

Upgrading Your Relationship Operating System

Once you've identified your bugs and understood your current code, you can systematically upgrade your relationship programming. This process is more challenging than other areas because relationships involve another person's programming interacting with yours, but it's absolutely possible.

Week 1: Pattern Recognition and Documentation

Spend this week simply observing your relationship code in action without trying to change it. Notice what triggers your programming, what situations activate anxiety, avoidance, people-pleasing, or other patterns. Document your automatic responses in relationships, what do you think, feel, and do when code is running? Notice the results your programming generates, what outcomes does your code create?

Write all of this down with curiosity rather than judgment. You're debugging, not criticizing yourself. Your code developed for good reasons, it protected you when you needed protection. It's just outdated now for creating healthy adult relationships.

By week's end, you should have clear documentation of your active relationship code, when it runs, what it does, and what results it creates.

Week 2: Trigger Identification and Response Design

This week, identify your top three relationship triggers, situations that reliably activate your old programming. For each trigger, design a new response that would come from upgraded code.

Example: Trigger: Partner wants a night alone. Old code response: Panic about abandonment, pursue for reassurance, create conflict. New code response: Recognize this is my anxiety code activating, not actual danger. Give space gracefully. Use the time for self-care. Trust that normal distance doesn't mean rejection.

Write out your new responses in detail. Make them specific enough that you could actually implement them. Then practice them mentally through visualization, imagine the trigger occurring and yourself executing the new response.

You're not trying to implement these yet in real situations, you're just designing and rehearsing upgraded code so it's available when you need it.

Week 3: Implementation in Low-Stakes Situations

Begin implementing your upgraded responses in lower-stakes situations. If your bug is people-pleasing, practice saying no to small requests. If it's rejection sensitivity, practice not pursuing when you feel the urge. If it's avoidance, practice staying present when you want to withdraw.

Start small. Don't try to completely rewrite your code in high-stakes situations immediately. Practice new programming in situations where the consequences of imperfect execution are minimal.

Document what happens when you run new code. What did you think and feel? How did others respond? What results did you get? What adjustments do you need to make?

This practice creates new neural pathways while building confidence that upgraded code actually works better than old code.

Week 4: Integration and Relationship Maintenance

This final week focuses on integrating your upgrades and establishing ongoing relationship maintenance practices.

Review the progress you've made over three weeks. What's improved? What's still challenging? What additional bugs need attention? Create a maintenance protocol for continuing to upgrade your relationship code beyond this initial month.

Most importantly, recognize that relationship code upgrade is ongoing work, not a one-time fix. Your programming developed over years or decades, it takes consistent practice to fully rewrite it.

The key is progress, not perfection. You don't need to completely eliminate old code immediately. You just need to catch it running more often, implement upgraded responses more consistently, and generate better results more frequently. Over time, upgraded code becomes default code.

Advanced Relationship Hacking Techniques

Once you've completed basic relationship code debugging, you can implement more sophisticated techniques for creating extraordinary relationships.

Conscious Partnership Programming

This advanced practice involves explicitly discussing relationship code with your partner and collaboratively debugging and upgrading together.

Most people never discuss their programming consciously. They just let their code and their partner's code interact automatically, creating whatever patterns emerge from that interaction. Conscious partners understand they're both running code, and they work together to upgrade their relationship operating system.

This requires vulnerability and maturity, but it creates dramatically better outcomes. Instead of blaming each other when patterns emerge, you identify code execution and debug together.

Example conversation: "I notice when you want space, my abandonment code activates and I want to pursue. I'm working on recognizing that's code running rather than actual danger. It would help me if you could reassure me that wanting space doesn't mean you're pulling away from the relationship, and I'll work on giving you space without making it mean something it doesn't."

This level of conscious communication allows both people to maintain awareness of code execution while implementing upgrades collaboratively. It's advanced practice, but it's incredibly powerful.

Marcus and his partner both had attachment code issues, his anxious, hers avoidant. Rather than letting these programs interact unconsciously (which would have been disastrous), they learned to identify when code was running and explicitly work with it.

When her avoidance code activated and she wanted distance, instead of just withdrawing (which would trigger his anxiety), she'd say: "I'm feeling my need-for-space code activating. This isn't about you or us, it's my programming. I need a few hours alone, but I'll check in after dinner."

When his anxiety code activated, instead of pursuing desperately (which would trigger her avoidance more intensely), he'd say: "My abandonment code is running right now. I'm working on not obeying it. I might need some reassurance that you're not actually leaving, but I'm trying not to be demanding about it."

This explicit code awareness allowed them both to get their needs met while not triggering each other's deepest fears. Their relationship worked because they understood they were two systems learning to interface well, not two people whose love should automatically override all programming.

Chapter 6: Career Simulation Cheats

Your career isn't determined by your education, skills, or even effort, it's the output of career code that's been running in your consciousness since you first formed beliefs about work, success, and your place in the professional world. This code generates opportunities, limitations, and results that match your deepest programming about what's possible for someone like you.

I discovered this during my transition from nursing school to building my own practice. Despite having excellent clinical training and genuinely caring about patients, my practice struggled for the first two years. I couldn't understand why, I was working harder than most other medical providers I knew, my patient outcomes were good, and I was constantly improving my skills. Yet I remained barely profitable, chronically stressed, and questioning whether I'd made a terrible mistake.

The breakthrough came during a conversation with a mentor who'd built a thriving practice seemingly effortlessly. I asked her secret, expecting advice about marketing or operations. Instead, she asked me a question that changed everything: "What do you believe about yourself as a business owner?"

I stumbled through some answers about being good with patients but not business-savvy, caring more about helping people than making money, not being the kind of person who was naturally successful at entrepreneurship. As I heard myself speak, I realized I was describing code, programming that said "people like me" weren't supposed to be good at business, that caring about financial success somehow meant not caring enough about patients, that my role was to be clinically excellent while someone else handled the business aspects.

This code was generating my reality as surely as my clinical knowledge was generating patient outcomes. I could have the best medical skills in the world, but if my career code said I wasn't supposed to succeed at business,

I wouldn't. The code would ensure it through unconscious self-sabotage, poor decisions, and opportunities I'd never notice or pursue.

The transformation began when I recognized my career struggles as code execution rather than inherent limitation. My difficulty building a successful practice wasn't a fixed characteristic, it was programming I could debug and rewrite.

The Career Programming Hierarchy

Most people's career programming falls into predictable levels, from most limited to most liberated. Understanding where you currently operate shows you exactly what needs upgrading.

Level 1: Survival Job Programming

At this level, work is purely about surviving, paying bills, not starving, maintaining basic security. There's no thought of fulfillment, growth, or meaning. Work is something you endure to fund life outside of work.

This programming says: Work is suffering you must endure. Jobs exist to pay bills, nothing more. Expecting fulfillment from work is naive. Your worth equals your paycheck. You're powerless to change your employment situation. You take what you can get and should be grateful for it.

People running this code often stay in jobs they hate for years or decades, feeling trapped by "golden handcuffs" or convinced they have no alternatives. They measure success by simply still being employed, not by whether work provides meaning or growth.

The telltale signs include constant complaining about work but taking no action to change it, feeling powerless to improve your employment situation, measuring worth entirely by external metrics like salary or title, having no vision for what work could be beyond surviving, and accepting mistreatment because "that's just how work is."

Robert exemplified this level. Despite having a college degree and valuable skills, he'd spent fifteen years in the same manufacturing job he hated, doing repetitive work that bored him, for a boss who treated him poorly. When friends suggested he look for something better, he'd say: "I'm lucky to have a job. The pay's okay. I've got benefits. I can't risk starting over somewhere new."

His programming wasn't about his actual situation, it was about what he believed was possible for someone like him. That code kept him trapped in limitation far beyond what circumstances required.

Level 2: Career Ladder Programming

At this level, you've moved beyond survival to thinking about advancement. Work is about climbing a ladder, earning promotions, achieving status, and progressing through established systems.

This programming says: Success means climbing corporate hierarchies. Follow the prescribed path and you'll advance. Work hard, prove yourself, earn promotions. Compete with others for limited positions. Measure success by titles, corner offices, and position on the org chart.

People running this code can achieve significant conventional success, good salaries, impressive titles, professional recognition. But they're still fundamentally operating within systems others created, following paths others designed, competing for opportunities others control.

The telltale signs include defining success by titles and position rather than fulfillment or impact, competing with colleagues for promotions, feeling anxious when others advance faster than you, following prescribed career paths without questioning whether they serve your actual goals, and measuring worth by where you are on established hierarchies.

Jennifer exemplified this level. As a marketing director at a Fortune 500 company, she'd successfully climbed from entry-level to senior leadership over twelve years. She earned $180,000, had an impressive title, managed a large team, and by conventional standards was highly successful.

But she was deeply unfulfilled. She'd sacrificed relationships, health, and personal interests to advance. She worked seventy-hour weeks. She competed constantly with peers. She lived in fear that someone younger and hungrier would surpass her. Despite appearing successful externally, she felt trapped in a game she no longer wanted to play but didn't know how to exit.

Her programming said career success meant climbing someone else's ladder. When I asked what success would look like if she designed the game herself, she looked at me blankly. "I don't understand the question. This is what success looks like, senior leadership at a major company."

She couldn't imagine success outside the framework her programming recognized. The code limited her vision to advancement within existing systems rather than creating her own.

Level 3: Achievement and Expertise Programming

At this level, you're focused on mastery, expertise, and being excellent at what you do. Work is about developing capabilities, creating value through your skills, and being recognized as highly competent.

This programming says: Excellence matters most. Become the best at what you do. Develop deep expertise. Measure success by the quality of your work and recognition from respected others. Mastery and craftsmanship are inherently valuable.

People running this code often become highly skilled professionals who love their work and take pride in doing it exceptionally well. They're craftspeople, experts, masters of their domains who derive satisfaction from excellence itself.

This is better than previous levels, at least you're doing work you care about and developing genuine expertise. But it still has limitations. You're often still trading time for money at high rates. You may be so focused on mastery that you miss opportunities for leverage or scaling. Your income is limited by your personal capacity.

The telltale signs include measuring worth by quality of work and expert recognition, pride in craftsmanship and mastery, reluctance to delegate because no one else can do it as well, income limited by your personal capacity to produce, and difficulty stepping back from hands-on work to build systems or teams.

Marcus exemplified this level as a master graphic designer. His work was exceptional, clients loved it, peers respected him, he'd won industry awards. He commanded premium rates and had more client requests than he could handle.

But he'd hit a ceiling. His income was limited by how many hours he could personally work. He couldn't take vacations without losing income. He struggled to hire help because training others felt like lowering quality standards. He was successful but exhausted, excellent but trapped by his own expertise.

His programming said success meant personal excellence. When I suggested building a team or creating systems that would let others execute his design approaches, he resisted: "But then it's not really my work. The quality won't be the same." His code couldn't conceive of success that didn't require his personal involvement in every project.

Level 4: Value Creation Programming

This is where real leverage begins. At this level, you're focused on creating value for others, solving problems, and building systems that generate results beyond your personal capacity.

This programming says: Success comes from value creation, not just personal excellence. Build systems that work without your constant involvement. Solve problems for many people simultaneously. Focus on results and impact, not just activity and effort. Leverage technology, processes, and other people's talents to multiply your contribution.

People running this code think in terms of scalability, systems, and leverage. They're not just excellent individual contributors, they're architects of value-creation systems. They understand that their role is designing and building rather than just executing.

The telltale signs include thinking about systems and leverage rather than just personal effort, building teams or processes that generate value without your direct involvement, focusing on results and impact rather than activity and hours, comfortable with others doing work differently than you would as long as results are good, and income that can scale beyond your personal time investment.

David exemplified this transition. As a software developer, he'd been excellent at his craft and well-paid for his skills (Level 3). But he hit the same ceiling Marcus faced, his income was limited by his hours.

The shift happened when he stopped thinking "How can I write more code?" and started thinking "What problems could I solve for multiple people with systems I build once?" He created a software tool that automated tasks for his target market. Instead of trading hours for dollars, he built something once that generated recurring value, and recurring revenue.

His income increased from $120,000 (limited by his hours) to over $400,000 (leveraged through a product that worked while he slept). More importantly, his role shifted from doer to designer, creating systems rather than just executing within them.

Level 5: Impact and Legacy Programming

This is the highest level, where work is primarily about creating meaningful impact and building something that matters beyond your own success. Money, recognition, and achievement are still important but secondary to contribution and legacy.

This programming says: My work should make a meaningful difference. Success means positive impact on others and the world. I'm building something that will continue creating value after I'm gone. My role is using my talents and resources to solve important problems. Wealth and success are tools for amplifying impact, not goals in themselves.

People running this code have transcended ego-driven achievement to focus on contribution. They're not working primarily for themselves anymore, they're working for something larger. Paradoxically, this often creates the greatest financial success, because they're solving real problems people genuinely value.

The telltale signs include primary focus on impact and contribution rather than personal gain, decisions based on what serves your mission rather than just what benefits you, willingness to take risks or make sacrifices for meaningful work, building systems designed to continue creating value beyond your involvement, and using success as a platform for solving bigger problems.

Dr. Sarah Chen exemplified this level. She'd built a successful medical practice (Level 4), but that wasn't enough. She used her success to create training programs for rural doctors, fund medical education for underserved communities, and develop healthcare models that could be replicated globally. Her work was about creating systemic change, not just running a good practice.

When I asked about her motivation, she said something that revealed her programming: "I realized I could have a successful practice and help a few hundred patients annually, or I could use that success as a foundation to impact thousands or millions through better systems and training. Once I saw that possibility, anything less felt like wasting the opportunity I'd been given."

That's legacy programming, success as a platform for contribution rather than an end in itself.

Identifying Your Career Code

Before you can hack your career programming, you need to see exactly what code is currently running. Most people have never consciously examined their career beliefs, they just follow paths that seem appropriate given their programming.

Take out a notebook and spend the next hour completing this career code audit. Write quickly and honestly without editing. You're debugging, not creating a resume.

Your Career Origin Story

What did you observe about work growing up? What did your parents say and demonstrate about careers and success? What messages did you receive about what was possible for people like you? What early experiences shaped your beliefs about work? What did you conclude about your own career potential? What were you told you should or shouldn't do professionally?

This traces your code back to installation points. Understanding where programming came from helps you recognize it as code rather than truth.

When I did this exercise, I remembered my mother working multiple low-wage jobs she hated, my father's sporadic employment and rage when money was tight, the constant message that "people like us" don't get good jobs, the pride I felt when I got into nursing school mixed with terror that I'd somehow been mistaken and didn't really belong there, and the deep belief that professional success was something other people achieved, people from stable backgrounds with connections and resources.

I could see clearly how that childhood programming was still running decades later, creating my struggles with building a practice despite having excellent clinical skills.

Your Career Beliefs Inventory

Complete these sentences quickly without overthinking:

Success in my field requires...

People who succeed in careers like mine are...

I'm the kind of person who...

My career ceiling is probably...

To advance, I would need to...

Making significant money requires...

The most I can realistically expect to earn is...

Truly meaningful work is...

Your automatic responses reveal your operating code. Look for patterns of limitation, permission, or possibility.

My responses revealed massive limiting code: Success requires connections I don't have. People who succeed have advantages I lack. I'm the kind of person who works hard but doesn't achieve major success. My career ceiling is probably comfortable middle-class professional. To advance, I would need to be more business-savvy and less idealistic. Making significant money requires compromising patient care or becoming corporate. Truly meaningful work doesn't pay well.

Every one of those beliefs was code generating my limited results. None of them were objective truth, they were programming creating my reality.

Your Career Pattern Analysis

Review your career history. For each position or phase, document: Why did you take this role? What did you hope to achieve? What patterns emerged? Why did you leave or stay? What opportunities did you pursue or avoid? What do you notice across your entire career history?

Most people discover they've been following a programmed script rather than making conscious choices aligned with their actual desires and capabilities.

My pattern was clear: I took safe, conventional paths. I avoided risks. I pursued approval from authority figures. I stayed in situations longer than I should have because security felt more important than growth or fulfillment. I repeatedly chose safety over possibility, approval over authenticity, convention over innovation.

This pattern wasn't random, it was code execution. My programming said: Play it safe. Follow prescribed paths. Don't take risks. Seek approval from gatekeepers. Stay in your lane. This code had served me in navigating from chaos to nursing school, but it was now preventing me from building the kind of practice I actually wanted.

Your Career Limitations Audit

What career moves have you considered but not pursued? What opportunities have you declined or avoided? What aspirations have you dismissed as unrealistic? What would you do if you knew you couldn't fail? What stops you from pursuing what you actually want?

The gap between what you want and what you pursue reveals your limiting code. The reasons you give for not pursuing possibilities are usually just your programming justifying itself.

I had a long list of things I'd considered but dismissed: Building a concierge practice focused on preventive care instead of conventional insurance-based medicine. Developing online patient education programs. Writing books about health. Creating training programs for other medical professionals. Launching a medical startup focused on patient experience.

Every one of these appealed to me deeply, and every one had been dismissed by my programming: That's too risky. You don't have business skills. You don't have the connections. That's not what nurses from your background do. Focus on being a good clinician and leave innovation to others.

My code was systematically filtering out opportunities that didn't match its programming about what was possible for someone like me.

Common Career Code Bugs

After years of working on my own career programming and observing others, I've identified recurring bugs that create predictable career limitations.

Bug #1: The Impostor Syndrome Virus

This bug constantly questions your qualifications and right to be in professional spaces, leading to self-doubt, over-preparation, and missed opportunities. It's especially virulent when your career success exceeds your family's historical achievements.

The code runs: If you're more successful than family precedent, trigger self-doubt. Question whether you really belong in professional spaces you occupy. Interpret success as luck or mistake rather than earned achievement. Over-prepare obsessively to prove you're qualified. Turn down opportunities because you feel you're not ready. Wait for someone to expose you as a fraud.

This bug is remarkably common among high achievers, especially first-generation professionals or those who've succeeded beyond their family's norm. The code creates constant anxiety that you'll be "found out" as not really belonging where you are.

Michael exemplified this perfectly. The first person in his family to graduate college, he'd earned an MBA from a top program and landed a consulting position at a prestigious firm. By objective standards, he was highly qualified and performing excellently. His reviews were consistently strong.

But he felt like a fraud constantly. He over-prepared obsessively for every client meeting, spending hours on presentations that colleagues completed in ninety minutes. He apologized frequently for his ideas even when they were good. He turned down opportunities to lead projects because he "wasn't ready yet" despite being more qualified than others who accepted similar roles.

His code was running perfectly: You don't really belong here. You got lucky. Real soon, someone will realize you don't actually deserve this position. Better to fly under the radar than risk exposure by being too visible or taking on too much responsibility.

The fix requires recognizing that feelings of impostorism are code executing, not accurate assessments of competence. Michael needed to install new programming: I earned my position through actual capabilities. My background is different from others here, but that's an asset not a liability. Discomfort in new situations means I'm growing, not that I don't belong. My perspective and experience are valuable precisely because they're different. Confidence comes from action, not from waiting until I feel ready.

He created a detailed competence inventory, objective documentation of his achievements, skills, positive feedback, and successful projects. When impostor code ran, he'd review this evidence instead of obeying the programming. He practiced attribution reframing, when he succeeded, he'd deliberately attribute it to his capabilities rather than to luck.

Within six months, he was promoted to senior consultant and was leading his own projects. The opportunities had always been there, his code had just prevented him from seeing himself as qualified to pursue them.

Bug #2: The Comfortable Ceiling Code

This bug creates an invisible upper limit on your career achievement, a ceiling that feels natural and appropriate but is actually just your programming's comfort zone.

The code says: Success beyond X level feels wrong or dangerous. If you approach that ceiling, trigger anxiety or self-sabotage. Create problems that bring you back to comfortable levels. Feel relief when you're back within familiar limits. This is as far as people like you get, and that's okay.

This ceiling varies by individual, for some it's $50,000 annually, for others $150,000, for others it's about titles or responsibility levels rather than income. But everyone has some level where their programming says "this is far enough."

The insidious part: the ceiling feels natural. It doesn't feel like programming, it feels like reality, like "this is just how far I can go" or "this is what's appropriate for someone like me."

Lisa discovered her ceiling at $85,000. Every time her income approached or exceeded that level, something would happen, unexpected expenses, poor business decisions, choosing low-paying opportunities over lucrative ones. She'd work her way back up to $85,000, approach $100,000, then something would knock her back down.

The pattern repeated so consistently it couldn't be coincidence. Her code had a ceiling, and when she exceeded it, self-sabotage protocols activated automatically to bring her back to comfortable levels.

The ceiling had been set by her family history, $85,000 was approximately what her parents had earned at their peak. Unconsciously, succeeding beyond that felt like betrayal, like becoming someone she wasn't, like losing her identity and connection to her roots.

The fix required recognizing the ceiling as programming rather than natural limit, and installing new code: My success honors rather than betrays my family. I can earn more while staying true to my values. Higher income doesn't change who I am fundamentally. My success creates opportunities for others. I'm expanding what's possible for my family line, not abandoning it.

She also needed to practice staying at higher income levels long enough for them to feel normal. When her income exceeded $85,000, instead of automatically self-sabotaging, she'd deliberately maintain that level for several months, allowing her nervous system to calibrate to the new baseline.

Within a year, she'd broken through to sustained income above $120,000. The ceiling had been real in its effects, but it was always just code that could be rewritten.

Bug #3: The Time-Worth Equation

This bug locks your sense of worth to hours worked, making it impossible to scale income beyond personal capacity.

The code says: Worth equals time invested. To earn more, work more hours. Earning without proportional time investment is wrong or invalid. My value is measured by how hard I work. If I'm not working long hours, I don't deserve success. Income should match time invested linearly.

This programming keeps you trapped in time-for-money models even when leverage opportunities exist. You literally can't imagine your worth being independent of hours worked.

James ran this code intensely as a consultant. He charged by the hour, worked extensive hours, and felt that was right and proper. When a colleague suggested he shift to value-based pricing or create products that didn't require his direct time, he had a visceral negative reaction.

"That feels like cheating," he said. "How can I charge for value I didn't directly create through my time? How can I earn money while I sleep? That doesn't seem right."

His code equated worth with time investment so completely that other models literally felt immoral. He couldn't conceive of his value being independent from hours he personally worked.

The fix required installing new programming: Value comes from results created, not time invested. Efficient solutions are more valuable than time-intensive ones. My expertise itself has value beyond each hour I work. Systems that generate value without my direct involvement are legitimate. My role is creating value, not just trading time.

James started shifting to project-based pricing focused on results. Instead of "$200 per hour," he'd price projects based on the value delivered: "This strategy typically increases revenue by $500,000 annually. My fee for designing and implementing it is $50,000."

Clients who understood value readily paid these prices. James discovered his income could increase dramatically while his working hours decreased, because he was being compensated for results and expertise rather than just time.

The bug had kept him trapped in artificial limitations that had nothing to do with his actual capabilities or the value he could create.

Bug #4: The Risk Aversion Override

This bug automatically rejects opportunities involving uncertainty, keeping you in safe but limiting situations.

The code runs: If opportunity involves uncertainty, automatically categorize as threat. Focus on what could go wrong rather than what could go right. Exaggerate risks and minimize potential rewards. Feel anxiety about change. Choose familiar limitations over uncertain possibilities. Justify risk avoidance with worst-case thinking.

This programming keeps people in jobs they hate, careers that don't fit, and situations far below their potential, because at least these situations are known and predictable.

Robert exemplified this bug. Despite being miserable in his manufacturing job for fifteen years, despite having skills that could translate to better opportunities, despite getting regular encouragement from friends to make a change, he stayed. His risk aversion code was too strong.

Every opportunity was immediately filtered through worst-case thinking: What if the new job doesn't work out? What if I can't handle it? What if I lose my benefits? What if I make a mistake and can't go back? What if the new company goes under? What if I'm starting over at my age and fail?

The code systematically generated anxiety about every possibility while downplaying the ongoing cost of staying stuck. It couldn't accurately assess risk because it was designed to avoid uncertainty, not evaluate opportunity.

The fix required installing new risk assessment code: Staying stuck has costs too, they're just more familiar so I notice them less. Uncertainty isn't the same as danger. I can handle challenges and course-correct. Most worst-case scenarios are unlikely and recoverable even if they occur. The risk of never trying might be greater than the risk of trying and adjusting.

Robert began taking tiny calculated risks to build his risk tolerance, applying for jobs he wouldn't take but wanted to practice interviewing for, taking on small freelance projects, learning new skills that could transfer to other roles. Each small risk he took successfully provided evidence that uncertainty didn't equal catastrophe.

Eventually, he accepted a position at a different company doing more interesting work for better pay. His risk aversion code hadn't disappeared, but he'd proven to himself that he could handle uncertainty and change, that his programming's dire predictions weren't accurate.

Bug #5: The Meaning-Money Dichotomy

This bug creates a false choice between meaningful work and financial success, making you believe you can have one or the other but not both.

The code says: Meaningful work doesn't pay well. Financial success requires compromising your values. If you care about making a difference, you shouldn't care about money. Charging appropriately for valuable work is greedy. People who earn significant money doing meaningful work have sold out somehow.

This bug is especially common in helping professions, education, nonprofits, and creative fields. It keeps talented people financially struggling while doing important work, because their code won't allow them to charge appropriately for genuine value.

Dr. Martinez exemplified this as a therapist. Her work was genuinely life-changing for her clients, she had exceptional skills, profound insights, and remarkable success rates. But she charged far below market rate because her code said charging more would mean she cared about money more than people.

She struggled financially, felt chronically stressed about money, couldn't invest in continuing education or better office space, and was heading toward burnout. Yet she felt proud of her low rates as proof of her dedication to helping people rather than making money.

Her code had created a false dichotomy: Either you care about helping people (and stay poor), or you care about money (and compromise your values). The possibility that charging appropriately would allow her to help more people better for longer literally didn't compute within her programming.

The fix required recognizing that the dichotomy was false: Money and meaning aren't opposed, money enables more sustainable, higher-quality meaningful work. Charging appropriately isn't greedy, it's valuing yourself and your work enough to sustain it long-term. People who pay appropriately value the work more and engage more seriously. Financial success doesn't require compromising values, it requires aligning your work with genuine value creation.

Dr. Martinez raised her rates to market level, expecting to lose clients. Instead, she found that clients who paid appropriate rates were more committed, followed through better, and achieved better outcomes. Her income increased by 60% while her client load decreased slightly, giving her more time and energy for each client. Her quality of work actually improved because she wasn't chronically stressed about money.

The meaningful work was always valuable. Her code had just prevented her from being appropriately compensated for that value.

The Career Code Upgrade Path

Once you've identified your bugs and understood your current programming, you can systematically upgrade toward mastery-level career code.

Week 1: Vision Design and Identity Upgrade

Most people's career struggles come from not having a clear vision of what success actually means to them, they're pursuing others' definitions while their own remains undefined.

Spend this week designing your actual career vision: What would meaningful success look like for you specifically? What impact do you want to have? What lifestyle do you want to support? What would you do if you knew you couldn't fail? What matters more than money? What level of income would genuinely feel abundant?

Write this out in detail. Not what you think you should want, but what you actually want when you're completely honest. Your vision might not match conventional success, and that's fine, but you need clarity about your actual target.

Then, identify the identity upgrade required: Who would you need to be to achieve that vision? What beliefs would that person hold? How would they make decisions? What risks would they take? How would they value themselves and their work?

Write out this upgraded professional identity in detail. You're designing the code that would generate your desired results.

Spend ten minutes each morning visualizing yourself as this upgraded professional identity. See yourself making decisions from that identity, experiencing the results that identity generates, and being the person who achieves your vision naturally.

Week 2: Opportunity Recognition Training

Your current code is filtering out opportunities that don't match its programming. This week focuses on retraining your pattern recognition to notice possibilities your old code dismisses.

Set an intention each morning: "I notice and consider opportunities today that align with my vision." This primes your brain to filter for possibilities.

Throughout the day, actively notice opportunities you would normally dismiss: Positions that seem "too senior" for you. Projects that seem "too ambitious." Collaborations that seem "not for someone like me." Business ideas that seem "too risky." Compensation levels that seem "too high for what I do."

When you notice yourself dismissing something, pause. Ask: "What if this is possible? What would it take? What would I need to believe or become to make this work?"

Write down at least three opportunities daily that you noticed and reconsidered instead of automatically dismissing. Then, choose one per week to explore further, even if just researching or having a conversation about it.

You're training your brain to see possibility where it currently sees impossibility.

Week 3: Value Recognition and Communication

This week focuses on recognizing your full value and learning to communicate it effectively without apologizing or minimizing.

Create a comprehensive value inventory: What problems do you solve? What results do you create? What transformations do you facilitate? What expertise do you have? What unique perspective do you bring? What would be lost if you weren't available?

Be specific and honest. Ask colleagues or clients what value you create, external perspective often reveals things you take for granted.

Then practice communicating value clearly: Write descriptions of what you do in terms of value rather than tasks. Practice explaining your worth without apologizing or qualifying. Role-play conversations about compensation where you confidently state your value.

Also begin pricing or positioning yourself based on value delivered rather than time invested or what feels "reasonable": "I help companies solve X problem, which typically creates $Y in value. My fee reflects that value creation." Or: "My expertise in Z area means I can deliver results others can't. My pricing reflects that unique capability."

This will feel uncomfortable if you're running undervaluation code. The discomfort is your old programming resisting upgrade. Keep practicing anyway.

Week 4: Strategic Risk Taking

This final week focuses on taking calculated career risks that your old code would have automatically rejected.

Identify three career moves that would advance your vision but feel risky: Applying for a position above your current level. Launching a side project or business. Proposing a significant idea at work. Asking for a substantial raise or promotion. Making a career change you've been considering. Pricing yourself at the value level you identified in week three.

For each risk, assess it accurately: What's the actual worst-case scenario? How likely is that scenario? How recoverable would it be if it occurred? What's the best-case scenario? What's the most likely scenario? What's the risk of not taking this risk?

Then, commit to taking at least one of these calculated risks this week. Take the action despite the discomfort. Your old code will scream warnings, that's expected. Take the action anyway.

Document what happens. Usually, the actual result is far better than your code predicted. This provides evidence that your risk aversion programming was creating false limitations.

Advanced Career Hacking Techniques

Once you've completed the basic four-week upgrade, you can implement more sophisticated techniques that accelerate career transformation.

The Income Multiplier Protocol

This technique systematically expands what income level feels natural and achievable to you.

Most people have an income "thermostat" set to a particular range. When you exceed it, you unconsciously create problems or expenses that bring you back down. When you fall below it, you unconsciously create opportunities that bring you back up. The goal is resetting this thermostat to a higher level.

Choose a target income that's 50-100% beyond your current level but not so far beyond it feels completely impossible. Perhaps if you currently earn $60,000, target $90,000-$120,000.

Spend five minutes daily visualizing earning this amount: See yourself receiving payment at this level. Imagine your bank account showing these numbers. Feel what it would feel like to earn this much. Experience the confidence, freedom, and ease of operating at this level.

Then, and this is crucial, start making decisions as if you already earned this amount: How would you allocate your time? What opportunities would you pursue? What would you say yes to, and no to? How would you price yourself? What investments would you make?

Begin making actual decisions based on this future income level, within reason. If someone making your target income wouldn't accept a certain project, don't accept it. If they would invest in particular training or tools, invest in them. You're training your system to operate from a higher baseline.

Marcus used this to shift from $35,000 to $240,000. He visualized earning $150,000 initially, then made business decisions as if he already earned that: pricing for $150,000-level value, declining low-value projects, positioning himself as premium, investing in business development. His income rose to meet his programming.

When he hit $150,000 and plateaued, he repeated the protocol targeting $250,000. His income followed his programming upward again.

The Expertise Positioning Strategy

This practice shifts how you position yourself from task-doer to expert, from employee-thinking to authority-thinking.

Most people position themselves by what they do: "I'm a graphic designer." "I'm a marketing coordinator." "I'm a software developer." This positioning is commodity-level, you're interchangeable with others who do the same tasks.

Expert positioning focuses on the unique value you provide: "I help tech startups create brands that attract ideal customers and funding." "I design marketing systems that reliably generate qualified leads." "I build software solutions for specific industry problems others can't solve."

This week, rewrite how you position yourself: From tasks to problems solved. From doer to expert. From "what I do" to "what value I create for whom." From commodity to specialist.

Then, practice introducing yourself from this expert positioning. Update your resume, LinkedIn, website, email signature. Communicate yourself as the expert you're becoming, even if it feels premature.

Positioning isn't lying, it's framing your actual capabilities in terms of value rather than tasks. You probably already solve the problems your expert positioning describes; you've just been positioning yourself as a task-doer rather than a problem-solver.

Jennifer made this shift from "marketing manager" to "customer acquisition strategist for B2B SaaS companies." Same skills, completely different positioning. The new positioning attracted higher-level opportunities and higher compensation because it communicated specialized expertise rather than generic execution.

The Strategic No Practice

One of the most powerful career hacks is learning to say no strategically, declining opportunities that don't serve your vision to create space for those that do.

Most people say yes to everything, staying constantly busy but never making real progress. They mistake activity for achievement, keeping themselves occupied with low-value work that prevents pursuing high-value opportunities.

This week, practice strategic no: Identify requests, opportunities, or commitments that don't align with your vision. Practice declining them gracefully but firmly. Notice the discomfort this creates, your programming likely says saying no is risky, rude, or will cost you opportunities. Say no anyway and observe what actually happens.

Also notice what saying no creates space for: Time for high-value activities. Energy for important projects. Availability when genuinely good opportunities arise. Focus on what matters most.

David discovered that saying no to small consulting projects (which kept him busy and brought in modest income) created space to develop a software product (which eventually generated far more income with less ongoing time investment). His code had kept him busy to feel secure, but that busyness was preventing the bigger opportunities his vision required.

Strategic no is advanced practice because it requires confidence that better opportunities will emerge if you create space for them. Your scarcity code will resist, insisting you should take everything offered. But abundance code knows that saying no to wrong opportunities makes space for right ones.

Your Career Transformation Timeline

Here's what successful career code upgrade looks like over twelve weeks:

Weeks 1-4: Foundation (Vision and Identity)

You're gaining clarity about what success actually means to you versus what you thought you should want. You're seeing your current limitations as code rather than fixed reality. You're identifying opportunities you would have dismissed before. You're beginning to position yourself differently.

By week four, you should have a clear career vision, identified your key limiting beliefs, begun noticing more opportunities, and taken at least one risk your old programming would have prevented.

Weeks 5-8: Development (Value and Visibility)

You're communicating your value more confidently. You're positioning yourself as expert rather than task-doer. You're pursuing opportunities that previously felt "not for me." You're pricing or negotiating from value rather than fear.

By week eight, you should see measurable progress, perhaps better opportunities emerging, higher compensation for existing work, more interesting projects, or advancement you'd been seeking. More importantly, you should feel different about your career, more confident, more strategic, more in control.

Weeks 9-12: Integration (Leverage and Scale)

Your upgraded career code is becoming automatic. You naturally recognize and pursue opportunities. You confidently communicate value. You make strategic decisions from your vision. You're building leverage, systems, reputation, expertise that multiplies your impact.

By week twelve, you should have achieved significant measurable results: 30-50% increase in compensation through various means, advancement in responsibility or position, new opportunities you wouldn't have accessed with old code, and most importantly, movement toward your actual career vision rather than just climbing ladders others built.

These aren't guarantees, results vary by starting point, industry, implementation consistency, and circumstances. But they represent typical outcomes when people systematically upgrade career code over three months.

Career Code Mastery

Ultimately, career code mastery means you're no longer limited by inherited programming about what's possible for someone from your background. You recognize opportunities others miss because their code filters them out. You create value others can't because your code allows you to think differently about contribution and impact.

You're running code designed for creation and contribution rather than just survival and climbing. You've shifted from being employed to being valuable, from following paths to creating them, from trading time for money to building leverage and systems.

The opportunities were always there. The potential was always in you. Your old code just filtered most of it out and kept you playing smaller games than you were capable of winning.

Upgrade the code, and suddenly you can see and access what was invisible before. Not because the world changed, but because your programming changed, which changed how you filter, interpret, and interact with the world.

Your career isn't determined by your credentials, skills, or even effort, it's determined by the code generating your professional reality. Change the code, change the career. It really is that straightforward.

Ready to rewrite your professional programming?

Chapter 7: Health System Override

Your health isn't determined by genetics, luck, or even lifestyle choices alone, it's the output of health code that's been running in your consciousness since you first formed beliefs about your body, vitality, and what's possible for your wellbeing. This code generates physical experiences that match your deepest programming about your body's capabilities, your worthiness of health, and what's "normal" for someone like you.

I discovered this during my residency when I started noticing a pattern that my medical training couldn't explain. I'd see two patients with virtually identical diagnoses, similar demographic profiles, and comparable treatment protocols. One would recover quickly and completely. The other would struggle, develop complications, and remain chronically unwell. The difference wasn't in their bodies, their treatment, or even their adherence to medical advice, it was in their consciousness, in the code they were running about health and healing.

This realization was initially disturbing. I'd been trained to see health as purely biological, genetics, pathogens, injuries, treatments. The idea that consciousness played a fundamental role in health outcomes felt dangerously close to "blaming the patient" or dismissing legitimate medical conditions as "all in your head."

But the evidence was undeniable. Patients who believed their bodies were strong and capable of healing generally healed better and faster. Patients who saw themselves as fragile or broken struggled regardless of treatment quality. Patients who expected problems often developed them. Patients who maintained positive expectations about recovery usually recovered better.

This wasn't wishful thinking creating miracles, it was programming affecting biology in measurable, predictable ways. The placebo effect alone proves that belief can create physical changes. Why wouldn't belief systems affect ongoing health, healing capacity, and vitality?

My own health journey confirmed this. Despite being a nurse practitioner who understood intellectually what constituted healthy lifestyle, I struggled with chronic fatigue, frequent illness, and a general sense that my body was unreliable and working against me. I was running health code installed during a chaotic childhood where my body was neglected, where illness meant burden, and where physical needs were subordinated to chaos management.

My health code said: Bodies are unreliable and require constant vigilance. Getting sick is inevitable and frequent. Energy is always limited. Pain and discomfort are normal. Taking care of my body is selfish when others need me. My body will break down eventually no matter what I do.

This programming was generating my physical experience as surely as my medical knowledge was generating my clinical decisions. I could prescribe perfect treatments for patients while my own health code sabotaged my wellbeing.

The transformation began when I recognized my health struggles as code execution rather than biological destiny. My chronic fatigue wasn't inevitable, it was the predictable output of programming that said bodies are fragile and energy is scarce. My frequent illness wasn't bad luck, it was the result of stress programming that kept my nervous system in constant fight-or-flight, suppressing immune function.

The Health Consciousness Hierarchy

Most people's health programming falls into predictable levels, from most limited to most vital. Understanding where you currently operate shows you what needs upgrading.

Level 1: Victim Consciousness

At this level, you experience your body as something that happens to you, unpredictable, unreliable, and fundamentally out of your control. Health is luck, genetics, or fate. You're passive in relation to your body's state.

This programming says: My body betrays me unpredictably. Health is genetic lottery, some people are healthy, others aren't, and there's not much you can do about it. Getting sick is random bad luck. My body is fragile and prone to breaking down. Medical professionals fix me when I break,

I'm passive in the process. Pain and illness are inevitable parts of life I must endure.

People running this code experience themselves as victims of their bodies. They don't see connection between lifestyle choices and health outcomes. They don't believe their actions significantly impact their wellbeing. They wait until things break, then seek external fixes.

The telltale signs include constant complaints about health without taking action to improve it, belief that health is determined by factors beyond your control, waiting until problems are severe before addressing them, relying entirely on external authorities to "fix" you, and having no sense of agency regarding your physical wellbeing.

Margaret exemplified this level. At fifty-three, she had multiple chronic health issues, diabetes, high blood pressure, chronic pain, frequent illness. She was on six medications, saw multiple specialists, and felt her body was constantly betraying her.

When her doctor suggested lifestyle changes, diet improvements, exercise, stress management, she'd say: "My body is just prone to these problems. My mother had them, my grandmother had them. It's genetic. There's not much I can do except take the medications."

Her programming literally prevented her from seeing that her choices affected outcomes. She was running pure victim code, health happened to her, and her role was passively receiving treatment rather than actively creating wellbeing.

Level 2: Management Consciousness

At this level, you recognize that your choices affect your health, but your focus is primarily on managing problems and avoiding illness rather than creating vitality.

This programming says: I need to be vigilant about health problems. Follow medical advice to manage conditions. Avoid things that might make me sick. Focus on not getting worse rather than getting dramatically better. Health is about managing problems, not creating vitality. Prevention means avoiding bad things, not building capacity.

People running this code are more active than victim consciousness, but still fundamentally reactive. They manage conditions rather than optimizing function. They avoid risks rather than building resilience. They're focused on not getting sick rather than becoming vibrantly healthy.

The telltale signs include focus on disease prevention rather than vitality creation, managing symptoms rather than addressing root causes, fear-based health decisions focused on avoiding problems, compliance with medical advice without questioning or optimizing, and measuring health by absence of problems rather than presence of vitality.

Robert exemplified this level. He followed his doctor's advice carefully, took his medications as prescribed, got regular checkups, avoided foods his doctor warned against. He was managing his health conditions adequately.

But he had no vision for genuine vitality. His goal was simply avoiding getting worse. He never exercised because he "didn't have health problems that required it yet." He ate adequately but not optimally. He managed stress but didn't build resilience. He was maintaining rather than optimizing, managing rather than thriving.

His programming said health meant not being sick, which kept him in a gray zone between illness and vitality, never terrible, never great, just managing along.

Level 3: Responsibility Consciousness

At this level, you recognize that you're largely responsible for your health outcomes and actively make choices to support wellbeing. You're proactive rather than reactive, creating health rather than just managing problems.

This programming says: My choices significantly affect my health. I'm responsible for my body's wellbeing. I can research and implement strategies for optimization. Prevention means building strength and resilience, not just avoiding problems. Health is created through consistent choices, not just lucky genetics.

People running this code take active responsibility for their health. They educate themselves, make conscious lifestyle choices, and see themselves as the primary agent of their wellbeing with medical professionals as advisors and partners.

The telltale signs include active research and implementation of health strategies, focus on building capacity rather than just avoiding problems, taking responsibility for outcomes rather than blaming external factors, willingness to experiment and find what works for your body, and measuring health by energy, capability, and vitality rather than just absence of disease.

Jennifer exemplified this level. When she developed early signs of insulin resistance in her late thirties, instead of just accepting medication and management, she researched nutrition, implemented dietary changes, started exercising regularly, improved her sleep, and managed stress. Within six months, her markers were completely normal without medication.

She wasn't lucky or genetically blessed, she took responsibility and implemented changes. Her programming said: "My body responds to how I treat it. I'm responsible for creating the conditions for health. I have significant control over my outcomes."

This level is far better than victim or management consciousness, but it still has limitations. It's still somewhat effortful, you're working to create health rather than health being your natural state. You may still see your body somewhat adversarially, as something requiring management and control.

Level 4: Partnership Consciousness

This is where real transformation occurs. At this level, you experience your body as an intelligent partner rather than a machine requiring management. You listen to its signals, trust its wisdom, and work with it collaboratively rather than controlling it.

This programming says: My body has innate intelligence and healing capacity. Physical symptoms are information and feedback, not just problems. I can trust my body's signals when I learn to listen. Health is a conversation between consciousness and biology. Optimization comes from partnership, not force or control. My body wants to be healthy and knows how, my role is supporting that natural drive.

People running this code have fundamentally different relationships with their bodies. They listen rather than override, trust rather than control, collaborate rather than force. They see their body as wise rather than problematic.

The telltale signs include ability to read and respond to your body's signals, trust in your body's innate healing capacity, comfortable relationship with your physical self, viewing symptoms as information rather than just problems to suppress, and collaboration with your body rather than fighting it.

Dr. Sarah Chen exemplified this level. She maintained remarkable health and vitality despite demanding professional responsibilities. When I asked her secret, she said something that revealed her programming: "I listen to my body and trust it. When I'm tired, I rest. When I need movement, I move. When I need certain foods, I eat them. When something feels off, I pay attention before it becomes a problem. My body tells me what it needs, I just had to learn to listen."

That's partnership consciousness, body and mind working together rather than mind controlling body.

Level 5: Integration Consciousness

This is the highest level, where the distinction between body and mind dissolves into unified consciousness. Health isn't something you create or manage, it's your natural state when you're aligned with your authentic self and living congruently.

This programming says: Body, mind, emotions, and spirit are one integrated system. Health arises naturally from alignment and congruence. Illness often reflects disharmony, living incongruently, suppressing authentic self, or being out of alignment with values and purpose. Healing means wholeness, bringing all aspects into harmony. Optimal health is my natural state when obstacles are removed.

People operating at this level often exhibit health that seems almost supernatural, rapid healing, resistance to illness, sustained vitality, and age-defying capacity. But it's not supernatural, it's what becomes possible when programming no longer interferes with the body's natural intelligence.

The telltale signs include health as natural byproduct of alignment rather than effortful creation, rapid healing and strong resilience, mind-body-spirit integration in all aspects of life, viewing illness as messenger about incongruence or needed changes, and vitality that seems effortless because it flows from wholeness.

Master teachers, certain long-lived indigenous people, and rare individuals who seem to transcend normal aging operate from this level. It's not that they have special biology, they've removed the programming that interferes with the body's natural capacity for health and vitality.

Identifying Your Health Code

Before you can hack your health programming, you need to see exactly what code is currently running. Most people have never consciously examined their health beliefs, they just experience whatever their programming generates and assume that's reality.

Take out your notebook and complete this health code audit. Be honest about what you actually believe, not what you think you should believe.

Your Health Origin Story

What did you observe about health and bodies growing up? How did your family talk about health, illness, and aging? What health problems did family members have? What messages did you receive about your own body? What significant health experiences shaped your beliefs? What did you conclude about bodies and health generally? What did you learn about your body specifically?

This traces your code back to installation points. Understanding where programming came from helps you recognize it as code rather than truth.

When I did this exercise, I remembered my mother's constant health complaints and her belief that "our family just has bad health." I remembered being sick frequently as a child and feeling like my body was unreliable. I remembered learning that taking care of your body was selfish when others needed you. I remembered my grandmother aging rapidly and painfully, creating the belief that bodies inevitably break down and betray you.

I could see how that childhood programming was still running decades later, creating health struggles despite my medical knowledge and healthy lifestyle efforts.

Your Body Beliefs Inventory

Complete these sentences quickly without overthinking:

My body is...

As I age, my body will...

Getting sick means...

Pain and illness are...

My body's natural state is...

Taking care of my body is...

Bodies are designed to...

My health is determined by...

Your automatic responses reveal your operating code. Look for patterns of trust or distrust, capacity or limitation, partnership or adversarialism.

My responses revealed deeply limiting code: My body is fragile and unreliable. As I age, my body will break down and limit me. Getting sick means my body failed me. Pain and illness are inevitable problems to endure. My body's natural state is vulnerable to problems. Taking care of my body is work I have to do. Bodies are designed to eventually fail. My health is determined by genetics and luck.

Every one of those beliefs was code generating my health experience. None were objective truth, they were programming creating my physical reality.

Your Health Pattern Analysis

Review your health history. What recurring patterns do you notice? What conditions appear repeatedly? What symptoms show up under stress? How quickly do you typically heal? How much energy do you generally have? What's your relationship with your body like?

Most people discover they've been running the same health patterns for years or decades, not because their body is fixed that way, but because their programming keeps generating the same experiences.

My pattern was clear: chronic low-grade illness, frequent minor infections, persistent fatigue regardless of sleep, slow healing from injuries, autoimmune flare-ups during stress. This wasn't random, it was code execution. My stress programming kept my nervous system in fight-or-flight, suppressing immune function and healing capacity. My unworthiness code made self-care feel selfish. My scarcity code said energy was always limited.

Your Health Limitations Audit

What level of health have you dismissed as unrealistic for you? What physical capabilities do you assume are impossible? What would you love to experience physically that your code says isn't available to someone like you? What physical goals have you abandoned as unattainable?

The gap between what you want and what you believe possible reveals your limiting code.

I had a long list: sustained high energy throughout the day, rarely getting sick, rapid healing from any injury or illness, feeling good in my body rather than constantly managing discomfort, exercising feeling energizing rather than depleting, sleeping well without effort, eating healthily feeling natural rather than restrictive, aging vibrantly rather than declining.

Every one of these felt unrealistic given my "natural" health patterns. But those patterns weren't natural, they were code execution. My programming was systematically filtering out possibilities that didn't match its beliefs about what was normal for someone like me.

Common Health Code Bugs

After years of working on my own health programming and observing countless patients, I've identified recurring bugs that create predictable health problems.

Bug #1: The Genetic Determinism Trap

This bug makes you believe your family's health history determines your destiny, creating fatalistic acceptance of inherited patterns.

The code says: Health problems run in my family, so I'll probably get them too. Genetics determine health outcomes more than choices. Fighting genetic destiny is futile. If my parents had certain conditions, I probably will too. My body is predetermined by my DNA to develop specific problems.

This programming creates self-fulfilling prophecies. You expect to develop family patterns, so you don't prevent them. You see early signs as confirmation of genetic destiny rather than opportunities for intervention. You feel powerless to change inherited trajectories.

The reality: while genetics influence health, they're far from deterministic. Epigenetics research shows that gene expression is heavily influenced by environment, lifestyle, beliefs, and behaviors. You can literally turn genes on or off through how you live and what you believe.

Michael exemplified this bug. His father had died of heart disease at fifty-five. His grandfather had died of heart disease at fifty-three. By forty, Michael was showing early signs, high blood pressure, concerning cholesterol levels, pre-diabetes.

His response: resignation. "Heart disease runs in my family. I'll probably go the same way. Not much I can do about genetics." He took medications but made minimal lifestyle changes, feeling it was somewhat pointless given his genetic destiny.

His programming prevented him from seeing that his father and grandfather had also shared lifestyle patterns, high stress, poor diet, no exercise, chronic sleep deprivation. The "genetic" pattern might have been more about inherited lifestyle and beliefs than DNA.

The fix required installing new programming: Genetics are influences, not destiny. Gene expression depends on environment and choices. I can prevent or delay conditions that run in my family through lifestyle. Early signs are opportunities for intervention, not confirmations of fate. My choices matter more than my DNA in determining outcomes.

Michael began treating early signs as warning signals rather than genetic inevitability. He made substantial lifestyle changes, diet, exercise, stress management, sleep optimization. Five years later, his markers were completely normal. He'd overridden his "genetic destiny" through upgraded programming and consistent action.

Bug #2: The Stress Normalization Protocol

This bug treats chronic stress as normal and unavoidable, preventing you from recognizing its health impacts or taking action to reduce it.

The code says: High stress is just part of modern life. Everyone is stressed, it's normal. I can handle it, stress doesn't really affect me. Stress is necessary for success and achievement. Relaxing or reducing stress means I'm lazy or not ambitious enough. My body should function well regardless of stress levels.

This programming keeps you in chronic fight-or-flight activation, which systematically suppresses immune function, impairs healing, disrupts sleep, reduces energy, increases inflammation, accelerates aging, and contributes to virtually every chronic disease.

The insidious part: your body adapts to chronic stress by numbing you to its signals. You stop noticing you're stressed because it's your baseline. You think you're fine while your body is slowly breaking down.

Lisa ran this code perfectly. As a high-achieving executive, she wore her ability to handle stress as a badge of honor. She worked twelve-hour days, responded to emails at all hours, took no real vacations, and prided herself on never slowing down.

She also had chronic health problems, frequent illness, digestive issues, insomnia, anxiety, hormonal disruption. She treated each symptom independently, never connecting them to chronic stress because her code said stress was normal and she was handling it fine.

The wake-up call came with a serious health crisis at forty-two, her body essentially forced her to stop through collapse. Only then did she recognize that her "normal" stress levels had been destroying her health progressively for years.

The fix required recognizing that chronic stress isn't normal or benign: Stress kills slowly but effectively. My body isn't designed for constant activation. Relaxation isn't laziness, it's essential for health. High performance requires recovery, not just effort. Stress management isn't optional, it's foundational for health.

Lisa completely restructured her life around stress reduction, shorter work hours, real vacations, daily practices for nervous system regulation, boundaries around availability, regular rest. Her health problems resolved systematically as her stress levels normalized. She discovered she could still be successful and achieve at high levels while actually taking care of her body.

Bug #3: The Self-Care Guilt Subroutine

This bug makes you feel guilty or selfish for prioritizing your health and physical needs, leading to chronic self-neglect.

The code says: Taking care of myself is selfish when others need me. I should put everyone else first and handle my own needs last. Spending time or money on my health is indulgent. Real giving means sacrificing your own wellbeing. If I'm too focused on my health, I'm being self-absorbed. My needs are less important than others' needs.

This programming is especially common in caregivers, helping professionals, parents, and people raised in environments where their needs were subordinated to dysfunction or others' demands.

Rachel exemplified this bug perfectly. As a therapist, mother of three, and primary caregiver for aging parents, she was constantly attending to everyone else's needs while completely neglecting her own. She couldn't remember the last time she'd exercised, eaten a proper meal sitting down, or slept more than six hours. She had multiple stress-related health problems but couldn't prioritize addressing them because her guilt code said taking care of herself meant neglecting others.

"I'll focus on my health when things calm down," she'd say. But things never calmed down because her code kept her taking on more responsibilities and putting herself last.

The reality: self-neglect doesn't serve anyone. When you're depleted, sick, and exhausted, you can't effectively care for others. Taking care of yourself first isn't selfish, it's necessary for sustainable giving. The oxygen mask principle applies: secure your own mask before helping others.

The fix required debugging her equation of self-care with selfishness: Taking care of myself allows me to serve others better. My wellbeing matters as much as others' wellbeing. Sustainable giving requires receiving and self-care. I model healthy boundaries by maintaining them. Neglecting myself doesn't help anyone, it just creates another person in need.

Rachel started with tiny acts of self-care that her code could tolerate, ten minutes of morning exercise, actually sitting down to eat lunch, going to bed thirty minutes earlier. As these practices improved her energy and capacity, her guilt decreased. She discovered she could actually be more present and helpful to others when she wasn't constantly depleted.

Bug #4: The Quick Fix Algorithm

This bug makes you seek immediate solutions to health problems while ignoring the underlying systems and patterns creating them.

The code says: Health problems should be fixed quickly. Take a pill, get a treatment, solve it fast. Symptoms are problems to eliminate, not information to understand. I don't have time for long-term lifestyle changes. Just fix this specific issue so I can get back to normal. Prevention and long-term health building take too long, deal with problems as they arise.

This programming keeps you in reactive mode, addressing symptoms without resolving root causes. You suppress one symptom only to have another emerge. You "fix" problems without changing the patterns creating them.

James ran this code intensely. Every health issue got a quick fix: Headaches? Take pills. Poor sleep? Sleep medication. Low energy? More caffeine. Digestive problems? Antacids. He had a pharmaceutical solution for every symptom but never addressed the underlying stress, poor diet, and lack of exercise creating all of them.

His quick fixes were generating side effects and new problems while the root causes continued undermining his health. He was suppressing symptoms without healing anything.

The fix required installing long-term thinking: Symptoms are information about underlying imbalances. Quick fixes often mask problems without resolving them. Real health comes from addressing root causes, not just symptoms. Prevention and system-building are more effective than reactive symptom management. Sustainable health requires lifestyle and pattern changes, not just treatments.

James began addressing root causes, stress management, dietary improvements, regular exercise, sleep optimization. The "quick" solutions took longer than popping pills, but they actually resolved problems rather than just suppressing them. Within six months, he no longer needed most of his medications because the underlying issues had been addressed.

Bug #5: The Body-Mind Separation Error

This bug treats body and mind as separate systems, preventing you from recognizing how deeply interconnected they are.

The code says: Physical health and mental/emotional state are separate. My thoughts and emotions don't significantly affect my physical health. Stress is "just mental," it doesn't really impact my body. Physical symptoms need physical treatments only. My body is like a machine that operates independently of my mental state.

This programming prevents you from recognizing how powerfully consciousness affects biology, and how physical health impacts mental and emotional states. You treat symptoms physically while ignoring psychological and emotional contributors.

The reality: every thought triggers neurochemical responses. Every emotion affects hormones. Chronic stress suppresses immune function. Depression often has physical components. Inflammation affects mood. The gut-brain axis influences both digestion and emotions. You cannot separate body and mind, they're one integrated system.

Michelle struggled with chronic pain that doctors couldn't explain or effectively treat. Every test came back normal, every treatment provided only temporary relief. She was frustrated that nothing "physical" was working.

When a doctor suggested the pain might be related to suppressed emotions and chronic stress, she resisted: "The pain is real. It's physical. This isn't psychological." Her code separated body and mind completely, preventing her from considering that real physical pain could have psychological components.

The breakthrough came when she finally tried mind-body approaches, therapy for processing trauma, stress reduction practices, emotional release work. The pain that had been resistant to physical treatments began resolving as she addressed the emotional and psychological components.

Her pain was real and physical. But it was being generated and maintained by mind-body processes, not just purely physical pathology. Once she debugged her body-mind separation error, she could access treatments that actually addressed root causes.

The fix required recognizing integration: Body and mind are one system. Thoughts and emotions directly affect physical health. Physical symptoms often have psychological components. Stress is physical, it creates measurable biological changes. Effective health approaches address the whole integrated system.

Upgrading Your Health Operating System

Once you've identified your bugs and understood your current programming, you can systematically upgrade toward vitality-level health code.

Week 1: Body Partnership Installation

Most health struggles come from an adversarial relationship with your body, seeing it as unreliable, broken, or working against you. This week focuses on shifting to partnership consciousness.

Each morning, spend five minutes in body awareness meditation: Simply notice your body without judgment. Notice sensations, feelings, energy levels. Thank your body for everything it's doing, heart beating, lungs breathing, immune system protecting, cells regenerating. Speak to your body as a trusted partner: "Thank you for all you do. I'm learning to listen to you better. We're in this together."

This might feel strange initially, your code might say bodies are machines, not conscious partners. Do it anyway. You're installing new programming that recognizes your body's intelligence.

Throughout the day, practice listening to your body's signals: What is it telling you it needs? Rest? Movement? Water? Different food? Pause before overriding its signals with caffeine, medications, or willpower. Ask: "What is my body trying to tell me?"

Document what you notice: When you listened to your body, what happened? When you overrode its signals, what resulted? What patterns emerge?

By week's end, you should notice improved connection with your body, increased awareness of its signals, and beginning recognition of its intelligence rather than seeing it as problematic machine.

Week 2: Vitality Expectation Upgrade

This week focuses on upgrading what level of health and energy you expect and accept as normal.

Most people's health expectations are calibrated to their family history, past experiences, and cultural norms that are far below optimal. You expect to feel "pretty good for your age" rather than genuinely vital.

Create a vitality vision: What would optimal health look like and feel like for you? Not "realistic" given current patterns, what would be amazing? How much energy? How good would you feel in your body? How quickly would you heal? How rarely would you get sick? How would you age?

Write this vision in detail. Then spend ten minutes daily visualizing yourself experiencing this level of vitality: Feel the energy in your body. Notice how good it feels to move. Experience the clarity and aliveness. Make it as real and visceral as possible.

Also begin documenting evidence of your body's capabilities: Times you heal quickly. Moments of high energy. Days you feel good. Physical capacities you have. Your body's strengths. You're training your brain to notice vitality rather than just problems.

When you catch yourself expecting limitation, "I'm tired because I'm getting older" or "I get sick every winter," challenge it: "What if that's just code? What if vitality is possible regardless of age or season?"

Week 3: Healing Capacity Recognition

This week focuses on recognizing and trusting your body's innate healing intelligence.

Your body heals cuts, fights infections, repairs injuries, regenerates cells, and maintains complex systems automatically without your conscious effort. This isn't luck, it's sophisticated biological intelligence that operates continuously.

Practice trusting this intelligence: When minor issues arise, before immediately treating them, pause. Ask: "What is my body trying to heal? What does it need from me to support that healing? How can I work with my body's intelligence rather than overriding it?"

This doesn't mean never seeking treatment, it means partnering with your body's healing capacity rather than assuming it's broken and needs external fixing.

Also research and understand your body's healing processes: How does immune function work? How do tissues repair? What supports healing? What interferes with it? Understanding your body's intelligence helps you trust and support it.

Document times you've healed: Injuries that resolved. Illnesses you recovered from. Times your body successfully handled challenges. Evidence of your body's healing capacity. This builds trust in its intelligence.

Week 4: Integration and Lifestyle Alignment

This final week focuses on aligning your lifestyle with your upgraded health consciousness.

Review your current lifestyle patterns: What supports the vitality you're creating? What undermines it? What changes would align your daily life with your health vision?

Make concrete changes in at least three areas:

Movement: Establish regular physical activity that feels good and energizing rather than punishing. Your code might say exercise is work, upgrade to "movement is celebration of what my body can do."

Nutrition: Improve your relationship with food from restriction or indulgence to nourishment. Eat in ways that actually make you feel good rather than following rigid rules or ignoring how food affects you.

Rest: Prioritize sleep and recovery. Your code might say rest is laziness, upgrade to "rest is essential for health and performance."

Stress: Establish daily practices for nervous system regulation, even five minutes of breathwork, meditation, or quiet reflection significantly impacts health.

Also establish ongoing health maintenance practices you'll continue beyond this month: Daily body awareness check-ins. Weekly vitality assessment. Monthly health optimization review. These prevent reverting to old programming.

Advanced Health Hacking Techniques

Once you've completed the basic four-week upgrade, you can implement more sophisticated techniques that dramatically accelerate health transformation.

The Energy Optimization Protocol

Most people accept their energy levels as fixed, "I'm just not a high-energy person" or "I'm always tired." But energy is programmable and can be systematically optimized.

Track your energy levels hourly for one week: Rate your energy 1-10 each hour. Note what you were doing before high-energy periods and before low-energy periods. Look for patterns in what increases versus depletes your energy.

Most people discover their energy isn't random, it responds predictably to food, activity, stress, sleep, and thoughts. Armed with this data, you can systematically optimize:

Identify your top three energy drains and minimize or eliminate them. Find your top three energy generators and increase them. Eat foods that increase energy for you rather than following generic advice. Time demanding activities for your high-energy windows. Manage your energy as carefully as you manage your time.

Marcus discovered that his afternoon energy crashes correlated with high-carb lunches, that his highest energy came after morning exercise, and that certain people and activities consistently drained him. By optimizing based on this data, protein-rich lunches, non-negotiable morning workouts, boundaries around energy-draining interactions, his baseline energy increased by 40%.

The energy was always available. His code and habits were just systematically depleting it.

The Inflammation Reduction Strategy

Chronic inflammation underlies most modern health problems, heart disease, diabetes, autoimmune conditions, cancer, cognitive decline, and accelerated aging. Reducing systemic inflammation dramatically improves overall health.

Implement these inflammation-reduction strategies:

Nutrition: Eliminate or dramatically reduce inflammatory foods (processed foods, excess sugar, industrial seed oils). Increase anti-inflammatory foods (wild fish, colorful vegetables, olive oil, nuts, berries). Many people feel dramatically better within weeks just from this shift.

Movement: Regular moderate exercise reduces inflammation. Sedentary lifestyle increases it. Even daily walking makes significant difference.

Sleep: Poor sleep increases inflammation dramatically. Optimize sleep quality and duration, this alone can transform health.

Stress: Chronic stress drives inflammation. Daily stress management practices reduce it measurably.

Connection: Social isolation increases inflammation. Meaningful relationships reduce it. Your social life affects your biology.

Jennifer implemented comprehensive inflammation reduction and within three months saw remarkable changes: Chronic joint pain resolved. Brain fog lifted. Energy increased substantially. Skin improved. She lost excess weight easily. Blood markers showed dramatic improvements.

Her doctors were impressed by "how well the medications were working." She hadn't changed medications, she'd reduced the inflammation that was causing most of her problems.

The Sleep Optimization System

Sleep is perhaps the most underrated health intervention. Poor sleep systematically undermines everything, cognitive function, emotional regulation, immune function, healing capacity, metabolism, hormones, and longevity.

Optimize your sleep through:

Environment: Cool, dark, quiet room. Comfortable bed. No screens or blue light before bed. Consistent sleep and wake times.

Wind-down routine: Establish calming pre-sleep ritual signaling your body it's time to rest. Hot bath, gentle stretching, reading, meditation, whatever works for you.

Timing: Most people need 7-9 hours. Find your optimal amount and protect it. Going to bed and waking at consistent times regulates circadian rhythm.

Supplements: Magnesium, glycine, or other sleep-supporting supplements if needed. But address root causes first.

Mind: If racing thoughts prevent sleep, establish a practice for clearing your mind, journaling, brain dump, meditation, progressive relaxation.

David struggled with insomnia for years, relying on medication. When he systematically optimized his sleep environment and practices, within two weeks he was sleeping deeply without medication. His energy, mood, and overall health improved dramatically.

The capacity for good sleep was always there. His environment and practices were preventing it.

The Biohacking Integration

This advanced practice involves using technology, tracking, and data to systematically optimize your biology.

Track key biomarkers: Get comprehensive blood work annually to track key health markers. Use wearable devices to monitor sleep, heart rate variability, activity. Track subjective measures, energy, mood, cognitive function, physical performance.

Experiment systematically: Change one variable at a time for at least two weeks. Track the effects objectively. Keep what works, discard what doesn't. Build a personalized optimization protocol based on your data.

Common effective biohacks include intermittent fasting for metabolic health and cellular repair, cold exposure for metabolism, immune function, and resilience, heat exposure through sauna for detoxification and cardiovascular health, red light therapy for cellular energy and healing, targeted supplementation based on your deficiencies, and breath work for nervous system regulation.

The key is systematic experimentation based on your data rather than following generic advice. What works for others might not work for you. Your biology is unique, optimize based on your responses.

The Mind-Body Medicine Protocol

This practice leverages the powerful mind-body connection for healing and health optimization.

Visualization for healing: Research shows that mental practice activates similar neural pathways as physical practice. Regularly visualize your body healing, immune system working optimally, tissues repairing, energy flowing. This isn't wishful thinking, it's communicating with your body through imagery.

Emotional release: Suppressed emotions often manifest as physical symptoms. Practices for emotional processing and release, therapy, journaling, somatic experiencing, emotional freedom technique, can resolve physical issues that resist purely physical treatment.

Meaning and purpose: Research consistently shows that people with strong sense of meaning and purpose live longer and healthier. Connecting with what matters to you isn't just psychological, it affects your biology.

Gratitude practice: Daily gratitude has been shown to improve immune function, reduce inflammation, enhance sleep, and increase overall wellbeing. Simple practice with measurable physical effects.

Social connection: Loneliness is as harmful to health as smoking. Strong relationships are as protective as any medical intervention. Prioritize meaningful connection, it's health medicine.

Your Health Transformation Timeline

Here's what successful health code upgrade looks like over twelve weeks:

Weeks 1-4: Foundation (Partnership and Awareness)

You're developing a different relationship with your body, partnership rather than adversarialism. You're noticing its signals rather than overriding them. You're beginning to trust its intelligence. You're establishing basic health practices.

By week four, you should notice improved body awareness, increased energy (even modest improvement is significant), better sleep quality, and most importantly, a felt sense that your body is capable rather than broken.

Weeks 5-8: Development (Vitality Building)

You're implementing substantial lifestyle optimizations based on your upgraded consciousness. You're eating differently, moving regularly, managing stress, sleeping better. Your body is responding, energy increasing, aches decreasing, mental clarity improving, resilience building.

By week eight, you should see measurable improvements: sustained energy throughout the day, significantly better sleep, reduced chronic symptoms, improved body composition, and quantifiable biomarker improvements if you're tracking them.

Weeks 9-12: Integration (Optimization and Mastery)

Your upgraded health consciousness is becoming automatic. Healthy choices feel natural rather than effortful. Your body feels like a trusted partner. You're operating from vitality rather than just managing problems. You're discovering capabilities you'd dismissed as impossible.

By week twelve, you should have achieved transformation: 40-60% increase in energy levels, resolution of chronic symptoms you'd accepted as normal, measurably improved biomarkers and health indicators, genuine vitality that others notice and comment on, and most importantly, completely different relationship with your body and health.

These aren't exaggerations, they're typical outcomes when people systematically upgrade health consciousness and align lifestyle with upgraded programming.

The key is recognizing that your health isn't fixed by genetics, age, or past patterns. It's generated by code that can be debugged and upgraded. Change the code, implement supporting practices, and your body responds, often dramatically and quickly.

Health Code Mastery

Ultimately, health code mastery means your body becomes a source of pleasure, vitality, and capability rather than a source of problems, limitations, and anxiety. You trust its intelligence, listen to its signals, and partner with it effectively.

You're no longer fighting your body or trying to force it to comply. You're collaborating with sophisticated biological intelligence that wants you to be healthy and knows how to get there if you provide proper conditions and remove interference.

The vitality was always possible. Your body always had remarkable healing and optimization capacity. Your old code just prevented you from accessing it, through stress keeping you in fight-or-flight, through self-neglect depleting resources, through poor lifestyle undermining function, through distrust preventing partnership.

Upgrade the code, align your lifestyle, and suddenly you can access health and vitality that felt impossible before. Not because your body changed fundamentally, but because you removed the programming that was interfering with its natural capacity.

Your health is programmable. Your vitality is upgradeable. Your body is far more capable than your current code allows you to experience.

Ready to rewrite your health programming and discover what your body can actually do?

PART 3: ADVANCED EXPLOITS

(Weeks 9-12)

Chapter 8: Emotional Operating System Manipulation

Your emotions aren't uncontrollable reactions to external events, they're the output of emotional programming that's been running in your consciousness since childhood, generating feeling states that match your deepest beliefs about safety, worthiness, and what's appropriate for you to experience. This programming operates so automatically that most people believe their emotions simply happen to them, when in reality they're the predictable output of code that can be identified, debugged, and rewritten.

I discovered this during a particularly dark period of my medical residency when I found myself experiencing the same emotional patterns I'd had since childhood, chronic anxiety, fear of abandonment, shame spirals triggered by minor mistakes, difficulty experiencing joy even during good moments, and an underlying emotional tone of "waiting for the other shoe to drop." These weren't just reactions to residency stress; they were code that had been installed decades earlier and continued executing regardless of circumstances.

The breakthrough came during a therapy session when my therapist asked a question that changed everything: "What if your emotions aren't happening to you? What if you're generating them through programming, and you could learn to generate different emotions by changing the code?"

This initially felt wrong, even offensive. My emotions were real, intense, and deeply felt. How could they be "just programming"? Wasn't that dismissing their validity or suggesting I was making them up?

But as I examined my emotional patterns more carefully, I saw the evidence: I had the same emotional responses to similar situations regardless of actual circumstances. I could predict my emotional reactions with remarkable accuracy. I had emotions that were disproportionate to triggers, tiny mistakes generating shame spirals, small slights creating intense hurt, minor uncertainties producing overwhelming anxiety. Most tellingly, I sometimes had emotions that contradicted objective reality, feeling unloved when surrounded by people who demonstrably loved me, feeling unsafe in genuinely safe environments, feeling worthless despite objective evidence of competence.

These patterns revealed programming, not just authentic responses to reality. My emotional code was executing automatically, generating feelings based on historical programming rather than present circumstances.

The Emotional Programming Hierarchy

Most people's emotional programming falls into predictable levels, from most reactive to most masterful. Understanding where you currently operate shows you what needs upgrading.

Level 1: Victim of Emotions

At this level, emotions feel like they happen to you, unpredictable forces you have no control over. You're at the mercy of your feelings, which seem to arise randomly and overwhelm you without warning.

This programming says: Emotions are things that happen to me. I have no control over what I feel. My emotions are caused by external events and other people. When I feel something, I must act on it. Strong emotions overwhelm me and prevent me from functioning. I can't help how I feel.

People running this code experience themselves as victims of their emotional states. They don't see connection between their thoughts and their feelings. They don't recognize patterns in their emotional responses. They feel powerless to change what they experience emotionally.

The telltale signs include feeling overwhelmed by emotions regularly, blaming others or circumstances for your emotional states, acting impulsively on emotions without pause, inability to function when experiencing strong feelings, and having no strategies for emotional regulation beyond waiting for feelings to pass.

Margaret exemplified this level. At forty-seven, she described herself as "just an emotional person," crying easily, getting angry unpredictably, feeling anxious most of the time, experiencing mood swings she couldn't anticipate or control. When asked what triggered different emotions, she'd say: "I don't know, they just happen. One minute I'm fine, the next I'm crying. I can't control it."

Her programming prevented her from seeing any connection between her thoughts, beliefs, and emotional experiences. Emotions felt like weather, external forces she could only endure, not influence or manage.

Level 2: Emotion Manager

At this level, you recognize that your emotions are somewhat manageable and you have some strategies for coping with them, but you're still primarily reactive, trying to control feelings through suppression or distraction.

This programming says: I need to control my emotions to function properly. Strong emotions are problems to be managed or suppressed. Certain emotions are acceptable, others aren't. I should think positive and avoid negative feelings. Emotional control means not showing or fully experiencing difficult emotions. Managing emotions means making them go away.

People running this code have moved beyond feeling completely victimized by emotions, but they're still fundamentally at war with their feelings. They're managing rather than mastering, controlling rather than understanding, suppressing rather than processing.

The telltale signs include using suppression or distraction to avoid difficult emotions, pride in "staying positive" or "not letting things bother you," difficulty identifying specific emotions beyond broad categories like "bad" or "stressed," physical symptoms from suppressed emotions, and emotional outbursts when suppression fails.

Robert exemplified this level. He prided himself on emotional control, "I don't let things get to me. I stay positive." But his "control" was actually suppression. He'd push down difficult emotions, distract himself when feelings arose, and maintain a positive facade regardless of what he was actually experiencing.

The cost: chronic tension, stress-related health problems, emotional numbness that prevented deep connection, and periodic breakdowns when suppressed emotions overwhelmed his control mechanisms. He was managing emotions, but at enormous cost and without actually processing anything.

Level 3: Emotional Awareness

At this level, you've developed the ability to recognize, name, and understand your emotions as they arise. You see connections between thoughts, beliefs, and feelings. You're no longer just reacting, you're observing your emotional experiences with some distance.

This programming says: I can observe my emotions without being completely controlled by them. Emotions provide information about my thoughts and needs. I can identify specific feelings rather than just broad categories. There's space between feeling and acting. All emotions are valid information, not problems to eliminate. Understanding my emotions helps me respond more effectively.

People running this code have developed metacognition about their emotional experiences. They can be angry while also observing that they're angry. They can feel anxious while recognizing the anxiety as an emotional state rather than objective danger. They've created space between stimulus and response.

The telltale signs include ability to name specific emotions accurately, understanding of your emotional triggers and patterns, pausing between feeling and acting, curiosity about emotions rather than just trying to eliminate them, and using emotions as information about needs and thoughts.

Jennifer exemplified this level. She'd learned to recognize her emotions as they arose, "I'm feeling anxious right now. That's interesting. What am I thinking that's creating anxiety?" She could feel difficult emotions without being overwhelmed or needing to suppress them. She used emotional information to understand herself better and respond more effectively.

This level is a significant improvement over reactivity or suppression. But it still has limitations, it's somewhat effortful, you're still somewhat identified with emotions even while observing them, and you may still feel at their mercy even if you understand them better.

Level 4: Emotional Mastery

This is where real power emerges. At this level, you recognize emotions as experiences you're generating through your programming, and you can deliberately influence what you generate through conscious thought and belief work.

This programming says: Emotions are generated by my thoughts and beliefs, not by external events. I can influence what I feel by changing how I think. Emotions are messengers and teachers, not just reactions. I can choose my emotional state in most circumstances. Difficult emotions can be processed and transformed. My emotional experience reflects my consciousness.

People running this code understand they're not victims of emotions, they're creators of emotional experiences through their programming. This doesn't mean they can instantly feel however they want, but they can systematically shift their emotional patterns through conscious work.

The telltale signs include ability to shift emotional states through conscious thought and practice, recognition that emotions reflect your programming rather than just circumstances, skillful processing of difficult emotions when they arise, capacity to maintain chosen emotional states even in challenging circumstances, and using emotions as feedback about beliefs that need examining.

David exemplified this level. When anxiety arose, he'd recognize: "I'm generating anxiety through catastrophic thinking. What am I believing that's creating this feeling?" He could then examine and challenge the thought, shifting the emotional experience. When anger emerged, he'd ask: "What boundary was crossed or what expectation was violated? What do I need to communicate?"

His emotions still arose, he wasn't suppressing or transcending them. But he recognized them as information generated by his programming, which gave him agency to process them skillfully and shift them when appropriate.

Level 5: Emotional Alchemy

This is the highest level, where you can consistently transform difficult emotions into useful energy, transmute suffering into wisdom, and maintain chosen emotional states regardless of circumstances. At this level, emotions become tools for creation rather than reactions to be managed.

This programming says: All emotional energy can be transmuted and used constructively. Suffering contains seeds of wisdom and growth. I can maintain my chosen emotional state regardless of external circumstances. Emotions are energy that can be directed consciously. My emotional state is my creation and responsibility. I use emotional experiences for evolution and contribution.

People operating at this level exhibit emotional mastery that seems almost superhuman, maintaining peace in chaos, finding opportunity in crisis, transforming pain into purpose. But it's not superhuman, it's what becomes possible when you fully understand emotions as programmable experiences.

The telltale signs include consistent ability to transform difficult emotions into growth and insight, maintenance of chosen emotional baseline regardless of circumstances, using emotional experiences to deepen wisdom and compassion, emotional resilience that strengthens through challenges, and emotions as fuel for creativity and contribution rather than just personal experience.

Master teachers, certain spiritual practitioners, and rare individuals who maintain peace and purpose through extreme challenges operate from this level. They haven't transcended emotions, they've mastered the art of emotional alchemy, transforming all experiences into fuel for growth and contribution.

Identifying Your Emotional Code

Before you can hack your emotional programming, you need to see exactly what code is currently running. Most people have never consciously examined their emotional patterns, they just feel what they feel and assume that's how emotions work.

Take out your notebook and complete this emotional code audit over the next week. This requires honest self-observation without judgment.

Your Emotional Origin Story

What emotions were acceptable in your family growing up? What emotions were forbidden or punished? How did your caregivers handle their own emotions? What did you learn about expressing feelings? What significant emotional experiences shaped your programming? What did you conclude about emotions generally? What did you learn about your own emotional nature specifically?

This traces your emotional code back to installation points. Understanding where programming came from helps you recognize it as code rather than inherent nature.

When I did this exercise, I remembered my mother's emotional volatility and my father's rage, learning that emotions were dangerous and unpredictable. I remembered being punished for expressing anger or sadness, learning that certain feelings were unacceptable. I remembered developing strategies for emotional suppression to stay safe. I remembered concluding that my emotions were too intense, too much, too wrong, that I needed to control them or they'd overwhelm me and others.

I could see how that childhood programming was still running decades later, creating my adult emotional struggles despite completely different circumstances.

Your Emotional Beliefs Inventory

Complete these sentences quickly without overthinking:

Emotions are...

When I feel strong emotions, I...

Anger means...

Sadness means...

Fear means...

Joy means...

My emotions are...

Other people's emotions are...

Your automatic responses reveal your operating code. Look for patterns of acceptance or resistance, trust or distrust, mastery or victimhood.

My responses revealed deeply problematic code: Emotions are dangerous and overwhelming. When I feel strong emotions, I try to suppress them or distract myself. Anger means I'm bad or out of control. Sadness means I'm weak. Fear means I'm in danger and need to escape. Joy means something bad will happen soon to balance it out. My emotions are too intense and wrong. Other people's emotions are my responsibility to manage.

Every one of those beliefs was code generating my emotional experience. None were objective truth, they were programming creating my reality.

Your Emotional Pattern Analysis

Review recurring emotional experiences in your life. What emotions do you experience most frequently? What triggers them predictably? What emotions do you rarely or never feel? What emotions are you most uncomfortable with? How quickly do you recover from difficult emotions? What's your emotional baseline, the state you default to?

Most people discover they have recurring emotional patterns that have been running for years or decades, not because situations keep happening to them, but because their programming keeps generating the same emotional responses.

My pattern was clear: chronic anxiety as my baseline, shame spirals triggered by mistakes or criticism, difficulty experiencing sustained joy or contentment, fear of abandonment activated by any perceived distance in relationships, suppressed anger that would build until it exploded, and an underlying emotional tone of "waiting for disaster." This wasn't random, it was code execution. My programming was systematically generating these experiences regardless of whether circumstances warranted them.

Your Emotional Avoidance Audit

What emotions do you avoid at all costs? What feelings are so uncomfortable you'll do almost anything to prevent experiencing them? What strategies do you use to avoid difficult emotions? What would happen if you fully felt what you're avoiding? What do you fear about certain emotional experiences?

The emotions you avoid reveal your deepest programming and often point to exactly what needs healing and upgrading.

I had a long list of avoided emotions: vulnerability (felt dangerous), deep grief (felt overwhelming), rage (felt unacceptable), genuine joy (felt like setting myself up for disappointment), peace (felt wrong, like I should be vigilant). My avoidance strategies included staying constantly busy, intellectualizing everything, helping others to avoid my own feelings, numbing through food or media, and maintaining control to prevent emotional intensity.

Common Emotional Code Bugs

After years of working on my own emotional programming and observing countless others, I've identified recurring bugs that create predictable emotional suffering.

Bug #1: The Emotion Suppression Protocol

This bug treats emotions as problems to be eliminated rather than information to be processed, leading to emotional numbness, physical tension, and eventual emotional explosions.

The code says: Strong emotions are dangerous or wrong. Push them down rather than feeling them. Stay busy or distracted to avoid emotional experiences. Think positively to override negative feelings. Don't let yourself fully feel difficult emotions. Expressing emotions means losing control. Emotional suppression equals strength.

This programming creates a pressure cooker, suppressed emotions don't disappear, they accumulate. Eventually they erupt in ways you can't control, or they manifest as physical symptoms, or they create chronic numbness that prevents feeling anything deeply.

Michael ran this code intensely. Raised to believe that "real men don't cry" and "emotions are weakness," he'd spent decades suppressing virtually all feelings. He prided himself on never getting emotional, staying rational, keeping control.

But the cost was enormous: chronic tension and headaches, difficulty connecting with his wife and children, emotional numbness that prevented feeling joy or love deeply, periodic explosive anger when suppression failed, and a sense of being dead inside while appearing fine outside.

The suppression that felt like strength was actually creating suffering and disconnection.

The fix required recognizing that emotions need processing, not suppression: Emotions are information and energy that need to flow through. Suppression creates problems, processing resolves them. Feeling emotions fully actually reduces their intensity and duration. Emotional expression is healthy, not weak. My body holds what my mind suppresses, physical symptoms often reflect suppressed emotions.

Michael learned practices for emotional processing, allowing himself to feel in safe contexts, naming emotions, expressing them through journaling or conversation, physical release through movement or breathwork. Initially terrifying (his code screamed that feeling would destroy him), he discovered that actually experiencing emotions wasn't nearly as dangerous as his programming predicted. In fact, processing them brought relief, connection, and aliveness he hadn't felt in decades.

Bug #2: The Emotional Perfectionism Trap

This bug expects you to feel positive or neutral all the time, treating negative emotions as failures that need fixing immediately.

The code says: I should be happy most of the time. Negative emotions mean something's wrong with me. If I'm feeling bad, I need to fix it immediately. Emotional struggles mean I'm failing at life. Other people don't have these problems as much as I do. Positive thinking should prevent negative feelings.

This programming creates a meta-problem, not only do you have difficult emotions, but you're also judging yourself for having them, creating additional suffering on top of the original emotion.

Lisa exemplified this bug. Whenever she felt sad, anxious, or angry, she'd immediately think: "What's wrong with me? Why can't I just be happy? I have so much to be grateful for, I shouldn't feel this way. I need to fix this feeling right now."

Her judgment about her emotions created more suffering than the emotions themselves. She was essentially having two problems: the original emotion plus harsh self-criticism for experiencing it.

The fix required recognizing that emotional diversity is normal and healthy: All humans experience a full range of emotions. Negative emotions aren't failures, they're part of normal human experience. Judging emotions creates more suffering than the emotions themselves. I can experience difficult emotions and be okay. Emotional ups and downs are natural, not signs of dysfunction.

Lisa learned to respond to difficult emotions with self-compassion rather than self-criticism: "Of course I feel sad right now, this situation is genuinely difficult. It's okay to feel what I feel." This simple shift, from judging emotions to accepting them, dramatically reduced her suffering.

Bug #3: The Emotional Dependency Algorithm

This bug makes your emotional state dependent on others' moods, approval, or behavior, preventing you from developing internal emotional stability.

The code says: My emotional wellbeing depends on how others treat me. If someone's unhappy with me, I feel terrible. I need others' approval to feel okay. Other people's moods affect my moods automatically. I can't be happy if people around me are unhappy. My emotional state is determined by external factors I can't control.

This programming keeps you emotionally reactive and dependent, constantly at the mercy of others' states and behaviors. You can't maintain chosen emotional states because your programming makes you absorb others' emotions or depend on their validation.

Rachel ran this code intensely as a therapist and people-pleaser. If a client struggled, she felt responsible and upset. If her husband seemed distant, she felt anxious and unworthy. If a friend didn't respond to texts quickly, she felt rejected. Her emotional state was constantly determined by others' behaviors and moods.

She was exhausted from emotional ping-ponging, feeling okay one moment, then upset because someone else seemed upset, then anxious because someone seemed distant, then guilty because she'd disappointed someone. She had virtually no internal emotional stability.

The fix required developing internal emotional independence: My emotional well-being is my responsibility, not others' responsibility. I can feel okay even when others are upset. Others' moods are their experience, not mine to absorb. Approval is nice but not necessary for my worthiness. I can maintain chosen emotional states regardless of others' behaviors.

Rachel established practices for emotional grounding, daily meditation, boundary setting, conscious choice of emotional state, and regular check-ins: "How do I actually feel about this situation, separate from how others feel?" She discovered she could maintain emotional stability and still be caring and responsive to others, she didn't need to absorb their emotions to be supportive.

Bug #4: The Emotional Authenticity Conflict

This bug creates internal struggle between your genuine feelings and what you think you should feel, leading to emotional confusion and inauthenticity.

The code says: I shouldn't feel this way, I should feel grateful/happy/content. My authentic emotions are wrong or inappropriate. Other people don't have these messy feelings. I need to feel what's appropriate for the situation. Authentic emotions might upset or burden others. I should be able to control what I feel.

This programming prevents you from even knowing what you genuinely feel because you're so busy judging those feelings and trying to feel something different.

David struggled with this acutely. When his father died, everyone expected him to be devastated. But he mainly felt relief, his father had been abusive, and David felt freed by his death. But his code said: "You should be sad. What's wrong with you that you feel relieved? This makes you a terrible person."

He couldn't be authentic about his feelings because they didn't match what he believed he should feel. He performed grief publicly while secretly feeling guilty about his relief. The internal conflict created more suffering than either emotion alone.

The fix required permission to feel authentically: All emotions are valid data about my experience. I don't have to feel what I "should" feel, only what I actually feel. Authentic emotions aren't right or wrong, they're information. Other people's expectations don't determine my legitimate feelings. Pretending to feel differently than I do creates disconnection from myself and others.

David learned to acknowledge his authentic feelings without judgment: "I feel relieved, and that's okay given my history with my father. I don't need to perform grief I don't feel. My authentic response is legitimate." This honesty, first with himself, then selectively with trusted others, brought relief and integrity.

Bug #5: The Trigger Hypersensitivity Code

This bug creates disproportionate emotional responses to minor triggers, as if your system is constantly on high alert for threats that may no longer exist.

The code says: Small slights mean major rejection. Minor criticism means I'm worthless. Any uncertainty means disaster is coming. People's neutral behaviors mean they're upset with me. I need to be hypervigilant to catch problems early. Better to overreact than miss a real threat.

This programming was often useful in genuinely threatening environments; it kept you safe by making you hypersensitive to danger signals. But in safe adult contexts, it creates constant false alarms, exhausting you with disproportionate emotional responses to non-threats.

Jennifer exemplified this. A coworker not saying hello in the hallway would trigger hours of anxiety: "She's mad at me. What did I do? She definitely hates me now." Her boss saying, "Can we talk later?" would create panic: "I'm getting fired. He found out I'm incompetent. This is it."

Her trigger sensitivity was exhausting; she lived in constant emotional reactivity to tiny ambiguous signals her code interpreted as catastrophic threats.

The fix required recalibrating threat assessment: Most triggers aren't actual threats; they're my code seeing danger that isn't there. People's neutral behaviors are usually about them, not about me. I can pause before reacting to assess whether this is actually dangerous. My code is hypersensitive from past legitimate threats, but I'm safe now. Not every trigger requires a response.

Jennifer learned to interrupt the trigger-reaction sequence: When triggered, pause. Ask: "Is this actually dangerous, or is this my code running?" "What would I think if I wasn't running hypersensitive programming?" "What's a more likely explanation than my catastrophic interpretation?"

Initially difficult (her code insisted every trigger was real danger), she eventually learned to distinguish false alarms from legitimate concerns. Her emotional reactivity decreased dramatically, along with her exhaustion.

The Emotional Code Upgrade Path

Once you've identified your bugs and understood your current programming, you can systematically upgrade toward emotional mastery.

Week 1: Emotional Awareness Development

Most emotional struggles come from unconscious reactivity, you're feeling before you know what you're feeling or why. This week focuses on developing conscious emotional awareness.

Practice emotional naming throughout each day: When you notice an emotional experience, pause and name it specifically. Not just "I feel bad" but "I feel anxious" or "disappointed" or "ashamed." The more specific, the better. Research shows that naming emotions ("affect labeling") actually reduces their intensity by engaging your prefrontal cortex.

Keep an emotion journal: Three times daily, write down what you're feeling and rate the intensity 1-10. Note what triggered the feeling if you can identify it. What thoughts accompanied the emotion? What physical sensations did you notice?

By week's end, you should have significantly improved emotional vocabulary and awareness, ability to identify specific emotions as they arise, recognition of your most common emotional patterns, and beginning understanding of connections between thoughts and feelings.

Week 2: Thought-Emotion Connection

This week focuses on recognizing how your thoughts and beliefs generate your emotional experiences.

Practice tracing emotions back to thoughts: When you experience an emotion, especially a difficult one, ask: "What was I thinking just before I felt this?" "What am I believing about this situation?" "What interpretation am I making that creates this feeling?"

You'll discover that specific thoughts create specific emotions with remarkable consistency. Thoughts about danger create anxiety. Thoughts about loss create sadness. Thoughts about violation create anger. Thoughts about inadequacy create shame.

Document thought-emotion patterns: "When I think _____, I feel _____." Build a personal database of your thought-emotion connections.

Also practice thought experiments: "If I believed something different about this situation, would I feel differently?" Usually yes. This proves emotions are generated by thoughts, not by situations directly.

By week's end, you should clearly see that your emotions are generated by your programming (thoughts and beliefs), not by external circumstances directly. This recognition is crucial, it shifts you from victim to creator of emotional experiences.

Week 3: Emotional Processing Practice

This week focuses on learning to process difficult emotions skillfully rather than suppressing or being overwhelmed by them.

When difficult emotions arise, practice this processing protocol:

Acknowledge: "I'm feeling [emotion]. That's okay, emotions are information."

Allow: Give yourself permission to feel it fully without trying to suppress or fix it immediately. Emotions that are allowed usually peak and subside within 90 seconds to a few minutes.

Explore: Ask curious questions. "What is this emotion trying to tell me?" "What need isn't being met?" "What boundary was crossed?" "What belief is creating this feeling?"

Express: Find healthy ways to release emotional energy, journaling, talking with trusted others, movement, breathwork, creative expression.

Integrate: "What can I learn from this emotional experience?" "How can this inform my choices moving forward?"

Practice this protocol with smaller emotions first, building capacity before applying it to intense emotional experiences.

Also practice emotional release techniques: Physical movement to release trapped emotional energy. Breathwork for emotional regulation. Journaling for processing. Talking with supportive others. Creative expression, art, music, writing.

By week's end, you should have established skillful practices for processing emotions rather than suppressing or being overwhelmed by them.

Week 4: Emotional Choice and Mastery

This final week focuses on developing ability to consciously influence your emotional state through deliberate thought and practice.

Practice emotional state shifting: Choose an emotion you'd like to experience more, perhaps peace, joy, confidence, or gratitude. Identify thoughts and beliefs that would generate that emotion. Practice thinking those thoughts deliberately. Notice how your emotional state shifts.

This isn't suppressing authentic emotions or fake positivity, it's learning that you can influence emotional experiences through conscious thought. You're developing agency rather than just reacting.

Also practice maintaining chosen emotional states during challenges: When situations arise that would normally trigger your old emotional patterns, consciously choose a different response. "My old code would generate anxiety right now. Instead, I'm choosing curiosity." "My default would be shame. Instead, I'm choosing self-compassion."

Establish daily emotional practice: Each morning, consciously choose your emotional baseline for the day. "Today I'm cultivating [peace/enthusiasm/confidence]." Throughout the day, notice when you drift from that chosen state and consciously return to it.

By week's end, you should have beginning mastery, ability to influence your emotional state through conscious choice, recognition that you're not a victim of emotions, and confidence that you can develop increasing emotional mastery through practice.

Advanced Emotional Hacking Techniques

Once you've completed the basic four-week upgrade, you can implement more sophisticated techniques that accelerate emotional mastery.

The Emotional Alchemy Practice

This advanced technique involves transforming difficult emotional energy into constructive energy for creation and contribution.

When you experience difficult emotions, anger, anxiety, grief, shame, instead of just processing and releasing them, practice transmutation:

Anger to Action: When anger arises, ask: "What boundary needs establishing?" "What needs to change?" "What action would address this?" Transform anger's energy into constructive change rather than destructive expression.

Anxiety to Preparation: When anxiety arises, ask: "What am I concerned about?" "What preparation would address this concern?" "What action would increase my readiness?" Transform anxiety into useful anticipation and preparation.

Grief to Gratitude: When grief arises, ask: "What did I love that I'm grieving?" "What am I grateful for having experienced?" Transform grief's energy into appreciation for what was rather than just pain about what's lost.

Shame to Growth: When shame arises, ask: "What can I learn from this?" "How can this experience help me grow?" "What does my response tell me about my values?" Transform shame into catalyst for development.

Marcus discovered he could use anxiety about upcoming presentations as energy for thorough preparation and passionate delivery. Lisa learned to transform grief about her divorce into deep appreciation for what she'd learned from the relationship. David transmuted anger about injustice into energy for creating positive change.

This practice doesn't deny or suppress emotions, it consciously directs their energy toward constructive ends.

The Emotional Baseline Optimization

Most people have an emotional baseline, a default state they return to when nothing particular is happening. For many, that baseline is mildly anxious, slightly dissatisfied, or vaguely uncomfortable. But baselines can be upgraded.

Identify your current emotional baseline: What do you feel when you're not feeling anything particular? What's your default emotional tone? Is it peace, anxiety, contentment, irritation, enthusiasm, boredom?

Choose your desired baseline: What emotional state would you like as your default? Peace? Contentment? Enthusiasm? Gratitude?

Practice installing new baseline: Spend 10-15 minutes daily experiencing your chosen baseline state through visualization, meditation, or embodied practice. Throughout the day, notice when you drift from your chosen baseline and consciously return to it. Over weeks and months, your default state will shift toward your chosen baseline.

Jennifer's baseline was chronic low-level anxiety, she'd been mildly anxious for so long it felt normal. She chose peace as her new baseline and spent three months practicing peace meditation daily and returning to peace throughout her day. Six months later, peace had become her actual default state, anxiety only arose in response to specific situations rather than being her constant companion.

The Meta-Emotion Practice

This advanced technique involves observing your emotional experiences with complete detachment while simultaneously feeling them fully, a paradoxical state that creates profound freedom.

Practice experiencing emotions while observing them: "I notice I'm feeling angry. This is interesting. The anger is intense. I'm not the anger, I'm the awareness experiencing anger. The anger arises and will subside. I'm the space in which anger appears."

This isn't suppression or dissociation, you're fully feeling while simultaneously maintaining awareness that you're larger than any particular emotion. You're the ocean experiencing waves, not just the waves.

This practice creates freedom because you realize you're not defined by or trapped in any emotional state. You can fully experience emotions without being overwhelmed because you recognize yourself as the space in which emotions arise rather than the emotions themselves.

Advanced practitioners can maintain this dual awareness consistently, fully engaged emotionally while simultaneously resting as the awareness that witnesses all emotional experiences.

The Emotional Resilience Building

This practice systematically builds capacity to handle increasingly intense emotions without being overwhelmed.

Start with manageable emotional challenges: Deliberately put yourself in situations that trigger moderate emotions, watching moving films, having meaningful conversations, engaging with art that evokes feeling. Practice feeling fully without suppressing or being overwhelmed.

Gradually increase intensity: As your capacity grows, engage with more intense emotional experiences. Process more difficult emotions. Sit with increasingly uncomfortable feelings without escaping.

This builds emotional resilience like physical exercise builds strength, progressive challenge with recovery creates increasing capacity.

Also practice emotional antifragility, learning to become stronger through emotional challenges rather than just surviving them: After difficult emotional experiences, consciously extract wisdom and growth. Use emotional challenges as development opportunities. Recognize that emotional struggles can increase your capacity and depth.

David systematically built emotional resilience by gradually engaging with emotions he'd suppressed for decades. He started with sad movies, progressed to therapy where he processed childhood trauma, eventually became able to sit with intense grief, rage, and vulnerability without being destroyed. His emotional capacity increased dramatically, he could feel more deeply while maintaining stability.

The Nervous-Excited Reframe: One of the Most Powerful Hacks

Here's one of the simplest yet most transformative emotional hacks: nervousness and excitement produce identical physical sensations. Your heart races. Your palms sweat. Your body trembles. Your breathing quickens. The physiological state is virtually indistinguishable.

The difference isn't in your body, it's in your programming.

When your code interprets these sensations as "nervousness," it generates avoidance. You want to escape, hide, postpone, cancel. Your nervous system interprets the arousal as danger and triggers your freeze or flight response. You stop. You retreat. You don't take the action.

When your code interprets the identical sensations as "excitement," it generates approach. You want to move forward, engage, take action. Your nervous system interprets the arousal as opportunity and triggers your activation response. You go. You advance. You perform.

Same body. Same sensations. Completely different outcomes based solely on the label you assign.

This is a fundamental reality hack: you can literally change what you're capable of doing by changing how you interpret your physiological state.

Marcus discovered this before a high-stakes client presentation. He'd been running nervous code his entire career, whenever important moments arrived, his body would activate with all the classic signs: racing heart, shaky hands, tight chest. His programming interpreted these signals as "I'm not ready. Something's wrong. I might fail." He'd feel paralyzed, his performance would suffer, and he'd confirm his belief that he wasn't good under pressure.

The night before a major pitch to a potential investor, Marcus decided to experiment. Instead of fighting the physical sensations or interpreting them as nervousness, he deliberately reframed them: "This isn't nervousness. This is excitement. My body is energized because this matters. This activation is exactly what I need to perform at my best."

He repeated this reframe several times, paying attention to the sensations without the old story attached. Heart racing? That's excitement, blood flowing to fuel peak performance. Palms sweating? That's excitement, my body preparing for action. Trembling? That's excitement, my nervous system activated and ready.

The next morning, walking into the presentation, the physical sensations returned. But this time, his code was running different programming. Instead of "I'm nervous and might fail," his internal dialogue was "I'm excited and ready." His body responded to this interpretation. His voice was steady. His thinking was sharp. His energy was magnetic. He closed the deal.

The only thing that changed was the label. The physiology was identical to every other time he'd felt "nervous." But the outcome was completely different because his programming had shifted from avoidance to approach.

This works because your brain doesn't distinguish between "nervousness" and "excitement" based on objective reality, it distinguishes based on your interpretation. When you tell yourself you're excited, your prefrontal cortex (the rational, capable part of your brain) stays engaged. When you tell yourself you're nervous, your amygdala (the threat-detection part) takes over and your higher functions diminish.

You can hack this. Before any challenging situation, presentations, difficult conversations, athletic performance, creative work, anything that activates your nervous system, pause and deliberately reframe: "I'm not nervous. I'm excited. My body is prepared. This activation is exactly what I need."

Say it out loud if possible. Feel it in your body. Notice the sensations without the old story. Your code will begin running different programming, and your performance will shift accordingly.

This is one of the most elegant hacks available: same input (physiological arousal), different interpretation (excited vs. nervous), completely different output (approach vs. avoidance, success vs. failure).

The simulation responds to your interpretation of reality, not to reality itself. Change the interpretation, change the outcome.

Your Emotional Transformation Timeline

Here's what successful emotional code upgrade looks like over twelve weeks:

Weeks 1-4: Foundation (Awareness and Processing)

You're developing conscious awareness of your emotional experiences instead of just unconsciously reacting. You're learning to name emotions specifically. You're seeing connections between thoughts and feelings. You're establishing practices for processing emotions skillfully.

By week four, you should notice reduced emotional reactivity, improved ability to identify what you're feeling, better understanding of your emotional patterns, and beginning skills for emotional processing rather than suppression.

Weeks 5-8: Development (Mastery and Choice)

You're implementing conscious emotional choice rather than just reacting automatically. You're transforming difficult emotions into constructive energy. You're maintaining chosen emotional states more consistently. Your emotional baseline is shifting toward your chosen state.

By week eight, you should see measurable improvements: significantly less emotional reactivity and drama, ability to shift emotional states consciously, reduced suffering from difficult emotions, improved emotional stability and resilience, and better relationships due to improved emotional regulation.

Weeks 9-12: Integration (Alchemy and Wisdom)

Your upgraded emotional code is becoming automatic. You naturally process emotions skillfully. You consistently transform challenges into growth. You maintain emotional mastery even during difficult circumstances. You're using emotions as tools for wisdom and contribution.

By week twelve, you should have achieved transformation: emotional experiences as information and energy rather than overwhelming forces, ability to maintain chosen states regardless of circumstances, wisdom extracted from emotional challenges, profound emotional resilience that strengthens through challenge, and emotional mastery that enhances all aspects of life.

These aren't exaggerations, they're typical outcomes when people systematically upgrade emotional programming and practice emotional skills consistently.

Emotional Code Mastery

Ultimately, emotional mastery means you're no longer a victim of your feelings. You recognize emotions as experiences you generate through your programming, information about your thoughts and needs, energy that can be directed consciously, and opportunities for growth and wisdom.

You're not suppressing emotions or transcending them, you're developing sophisticated relationship with them. You can feel deeply while maintaining stability. You can process difficult emotions without being destroyed. You can transform emotional energy into wisdom and contribution.

The capacity for emotional mastery was always available. Your old code just prevented you from accessing it, through suppression that created numbness, through reactivity that created chaos, through judgment that created meta-suffering, through dependency that prevented internal stability.

Upgrade the code, practice emotional skills, and suddenly you have access to emotional freedom and mastery that felt impossible before. Not because emotions changed fundamentally, but because your relationship with them transformed.

Your emotions are programmable. Your emotional experience is upgradeable. Your emotional mastery is available through conscious development.

Ready to transform your emotional operating system?

PART 4: BECOMING THE ADMIN

(Ongoing)

Chapter 9: The Master Programmer

After years of debugging, upgrading, and optimizing every system in my consciousness, I finally understood the deepest truth about reality programming: you are not just a user of your own system, you are the master programmer, the system administrator, and the ultimate authority over what code runs in your reality.

This realization changed everything. I stopped seeing myself as someone who had to work within the limitations of my programming and started seeing myself as the creator of that programming. I stopped trying to overcome my limitations and started rewriting the code that created those limitations. I stopped being a victim of my circumstances and started being the architect of my reality.

The journey from poverty to success, from chaos to stability, from survival to thriving, was ultimately a journey from unconscious programming to conscious creation. Every upgrade, every debug, every optimization was preparing me for this final understanding: I am the master programmer of my own existence.

This chapter is about making that transition, from someone who hacks their programming occasionally to someone who consciously programs their reality continuously. It's about moving from user-level access to admin-level control, from reactive debugging to proactive design, from fixing problems to creating possibilities.

Understanding Master Programmer Consciousness

Being a master programmer of your reality means understanding several fundamental truths that most people never fully grasp.

Truth 1: You Are the Author of Your Code

Every belief, habit, and automatic response in your system was either installed by you or accepted by you. Even programming installed during childhood was only integrated into your system because you, at some level, chose to accept it as true or necessary.

This doesn't mean you consciously chose limiting beliefs or that childhood trauma was your fault. It means that at every moment, you have the power to examine your code and decide whether to keep running it or rewrite it.

Most people live their entire lives running code they never consciously chose, inherited beliefs, cultural programming, childhood conditioning. They experience this code's outputs as "reality" without ever questioning whether different code might generate different realities.

Master programmers understand that all code is optional. Every belief can be examined. Every pattern can be modified. Every automatic response can be upgraded. Nothing in your programming is fixed unless you treat it as fixed.

When I discovered a belief creating limitation, "People like me don't achieve major success," I could see it clearly as code rather than truth. Code I could keep running or rewrite. The limitation existed only because I was executing that program. Stop running the code, install different code, and the limitation dissolves.

Truth 2: Your Code Is Always Running

Your programming executes continuously, every moment, generating thoughts, emotions, behaviors, and experiences whether you're paying attention or not. This means you're always programming yourself, either consciously or unconsciously.

Most people program themselves unconsciously, randomly reinforcing whatever patterns happen to be active. They think the same thoughts repeatedly, generating the same neural pathways. They react the same ways to similar situations, strengthening the same response patterns. They make the same types of choices, reinforcing the same decision algorithms.

This unconscious programming is why people's lives tend to be so consistent over time. Not because external circumstances are fixed, but because their code keeps executing the same programs, generating similar outputs year after year.

Master programmers take conscious control of this continuous programming process. They deliberately choose what code to run, consciously reinforce desired patterns, and systematically install upgrades. Instead of random programming happening to them, they're intentionally programming themselves toward desired outcomes.

I practice this through daily programming sessions, fifteen minutes each morning consciously installing and reinforcing the code I want running. I visualize myself operating from my chosen programming, I practice thoughts that align with my desired beliefs, I mentally rehearse behaviors that execute my upgraded code.

This daily practice ensures I'm programming myself consciously rather than letting random reinforcement shape my code.

Truth 3: The System Responds to Admin Commands

When you truly understand that you have admin access to your own consciousness, you can make changes that seem impossible from a user-level perspective. You can modify core programming that users can't touch. You can install new operating systems. You can create entirely new functionality.

The difference between user-level and admin-level access is profound. Users can only work within existing parameters, make surface changes, and use whatever features their system provides. Admins can modify the system itself, change core rules, and create new capabilities.

Most people operate at user level, they try to achieve different results while running the same underlying code. They work harder within their limitations rather than changing the code creating those limitations. They accept their programming as fixed and try to optimize within its constraints.

Master programmers operate at admin level, they modify the code itself. When they encounter limitations, they don't try to work around them; they rewrite the programming creating those limitations. When they want new capabilities, they don't hope for external solutions; they install new code that generates those capabilities.

The shift from user to admin consciousness happened for me when I stopped asking "How can I work harder within my limitations?" and started asking "What code is creating these limitations and how do I rewrite it?" That single shift in perspective unlocked transformation that years of hard work within my limitations never achieved.

Truth 4: Reality Is Your Output

What you experience as external reality is actually the output of your internal programming. Change the code, and the output changes predictably and consistently.

This doesn't mean you literally create physical reality with your thoughts, that's an oversimplification that leads to magical thinking. It means your programming determines what you notice, how you interpret it, what it means to you, how you respond, and therefore what results you create.

Two people can encounter the identical situation and have completely different experiences based on their programming. One person's code filters for opportunities and generates optimistic interpretations, creating positive outcomes. Another person's code filters for threats and generates pessimistic interpretations, creating negative outcomes. Same situation, different code, completely different realities experienced.

Master programmers understand that their experienced reality is their output. When they don't like their current reality, they don't primarily blame circumstances or try to force external changes. They examine their code and upgrade the programming generating unwanted outputs.

When my medical practice struggled initially, I spent months trying to fix external factors, marketing, operations, pricing. These helped marginally. The transformation happened when I recognized my struggling practice as output of my limiting code. Once I upgraded the programming, from "I'm not supposed to succeed at business" to "I create value and attract abundance," external results shifted dramatically without me working harder.

The circumstances hadn't been the problem. My code had been generating the struggling practice. Change the code, change the output.

Truth 5: You Are Responsible for Your System's Performance

As the master programmer, you're responsible for your code and its results. You can't blame bugs on the original programmers. You can't expect others to fix your code for you. You must maintain, update, and optimize your system continuously.

This responsibility might feel heavy initially. It's easier to blame circumstances, genetics, bad luck, or other people for your struggles. Taking full responsibility means accepting that your current results, however unsatisfying, are outputs of your current programming.

But this responsibility is also profoundly liberating. If you're responsible for your code, you have the power to change it. If your results are outputs of your programming, you can change your results by changing your code. Responsibility equals power.

Master programmers embrace this responsibility fully. They don't waste energy blaming external factors or waiting for others to fix their system. They take ownership of their code, their outputs, and their continuous development.

When my health was poor, I could have blamed genetics, bad luck, or my stressful profession. Taking responsibility meant recognizing that my health struggles were outputs of my stress programming, self-neglect code, and beliefs about bodies being unreliable. Once I accepted responsibility, I could upgrade the code and transform the outputs.

Responsibility isn't about guilt or self-blame for past problems. It's about claiming your power to create different results by running different code.

Master Programmer Capabilities

As you develop master programmer consciousness, you gain access to capabilities that most people never discover.

Capability 1: Real-Time Code Modification

Master programmers can adjust their programming based on immediate feedback and changing circumstances. They debug limiting beliefs as they arise, install new capabilities as they're needed, and optimize their programming continuously.

This real-time modification includes awareness algorithms that monitor your code execution continuously, pattern recognition systems that identify recurring problems quickly, instant debugging protocols that can identify and fix limiting beliefs in the moment, adaptive programming that can modify itself based on circumstances, and optimization subroutines that continuously improve performance.

I practice this constantly. When I notice limiting thoughts arising, "This is too hard" or "I can't handle this," I immediately recognize them as code and consciously choose different programming: "This is challenging, and I'm capable of meeting challenges" or "I'm learning and growing through difficulty."

This real-time debugging prevents limiting code from running unchallenged. Each time I catch and modify limiting programming in the moment, I weaken those old patterns and strengthen upgraded ones.

Capability 2: System Architecture Design

Master programmers understand how to create entire systems rather than just modifying individual programs. They design coherent programming architectures where all subsystems work together harmoniously.

This includes creating integrated systems where all subsystems support each other effectively, hierarchical programming that organizes code into logical layers, feedback loop integration that creates effective communication between different parts of consciousness, conflict resolution protocols that handle conflicts between different systems gracefully, and system coherence algorithms that maintain consistency across all programming.

When I upgraded my money code, I also had to upgrade my relationship code to be compatible, if my money programming said "charge premium prices" while my relationship code said "people-please and don't upset anyone," these systems would conflict. Master programmers design coherent systems where all parts work together.

I now think in terms of my entire system architecture: How does my money code interact with my purpose code? Does my health programming support my career code? Are my relationship algorithms aligned with my values? This systems thinking ensures all my programming works together rather than fighting itself.

Capability 3: Predictive Programming

Master programmers write code that anticipates future needs and challenges, creating systems that become more effective over time rather than just responding to current situations.

This includes scenario planning systems that anticipate future challenges and opportunities, adaptive learning algorithms that learn from experience and improve over time, proactive problem-solving that addresses potential issues before they become problems, growth trajectory programming that guides development toward desired outcomes, and environmental adaptation systems that adjust to changing circumstances automatically.

I practice this by regularly asking: "What challenges am I likely to face in the next three months? What programming upgrades would help me handle those challenges effectively? What capabilities will I need that I don't currently have?"

This predictive approach means I'm installing programming for future challenges before they arise, rather than reactively debugging after problems occur. I'm programming proactively rather than just debugging reactively.

Capability 4: Meta-Programming

This represents the highest level of programming mastery, writing code that modifies other code, creating self-improving systems that continuously upgrade themselves.

Meta-programming includes self-modifying algorithms that can rewrite themselves based on performance feedback, recursive improvement systems that create increasingly better versions of themselves, optimization compilers that optimize other programming for better performance, adaptive architecture that can restructure itself based on changing needs, and evolutionary algorithms that evolve toward better solutions over time.

This is advanced practice, but it's incredibly powerful. Instead of having to manually upgrade every piece of code, you create systems that upgrade themselves based on feedback and results.

I've installed meta-programming that automatically examines my beliefs when I get results I don't want: "This outcome suggests limiting code is running. What belief might be generating this result? How could I upgrade that belief?" This meta-level awareness means my system debugs itself rather than waiting for me to consciously identify every bug.

Capability 5: Legacy System Integration

Master programmers can preserve useful elements from old programming while upgrading the overall system. Not all old code is bad, some patterns that served you in the past may still be valuable if integrated properly.

This includes value extraction systems that identify useful elements in old programming, compatibility protocols that integrate old and new systems smoothly, gradual migration algorithms that transition from old to new systems without disrupting function, preservation subroutines that maintain important functionality during upgrades, and optimization integration that improves old systems while preserving their benefits.

My survival-oriented childhood programming, while limiting in many ways, also installed useful capabilities, resourcefulness, resilience, ability to function under stress, strategic thinking. Master programmers don't just delete all old code; they extract valuable elements and integrate them into upgraded systems.

I kept the resilience and resourcefulness from my old programming while removing the chronic anxiety and scarcity thinking. This selective preservation and integration creates more robust systems than completely deleting everything and starting over.

The Master Programmer's Toolkit

Master programmers develop and use sophisticated tools for consciousness programming that go beyond basic debugging and upgrading.

Tool 1: Advanced Consciousness Debugging

This includes awareness algorithms that monitor code execution in real-time, noticing what programming is running moment by moment. Pattern recognition systems that identify recurring bugs by analyzing thoughts, emotions, and behaviors over time, seeing patterns that aren't obvious in individual moments.

Root cause analysis protocols that trace problems back to their source code, understanding what core beliefs generate surface symptoms. Testing subroutines that evaluate the effectiveness of code modifications, measuring whether changes actually improve outcomes, and version control systems that track changes and allow rollbacks when modifications don't work as intended, maintaining records of what's been tried.

I maintain detailed consciousness logs, documentation of what code I'm running, what results it generates, what modifications I make, and what effects those modifications have. This systematic tracking allows me to debug more effectively because I can see patterns across time rather than just reacting to individual moments.

Tool 2: Belief Compilation Systems

These are tools for creating and installing new beliefs effectively. They include logical consistency checkers that ensure new beliefs integrate properly with existing code without creating conflicts or contradictions.

Emotional resonance generators that create positive emotional associations with new beliefs, making them feel true rather than fake. Behavioral evidence compilers that reinforce new beliefs through consistent actions and experiences, proving to your system that the new code works.

Integration protocols that make new beliefs feel natural and automatic rather than forced or artificial, and validation systems that confirm new beliefs are working as intended and producing desired results, providing feedback on effectiveness.

When I install new beliefs, I use this systematic approach. If I'm installing "I create value that others pay for generously," I check: Does this conflict with other beliefs? Does it feel emotionally resonant or fake? What actions can I take to prove this belief? How can I make this feel natural? How will I know if it's working?

This systematic approach to belief installation is far more effective than just repeating affirmations and hoping they stick.

Tool 3: Reality Generation Optimization

These are tools for maximizing the effectiveness of your reality-creation systems. They include outcome prediction algorithms that forecast the results of different programming choices, helping you choose optimal code.

Probability optimization systems that improve the likelihood of desired outcomes by running code that increases favorable probabilities. Resource allocation protocols that distribute energy and attention strategically rather than randomly.

Feedback loop analyzers that identify what's working and what needs adjustment, using results as data for optimization, and performance metrics that track progress toward desired outcomes, measuring whether your programming is actually generating the results you want.

I regularly assess: "What programming am I running? What results is it generating? Are those the results I want? If not, what code would generate better results? How can I install and run that code more effectively?"

This systematic optimization approach treats reality creation as an engineering problem with measurable inputs, processes, and outputs that can be continuously improved.

Tool 4: Consciousness Expansion Techniques

These tools systematically expand your awareness and capabilities beyond current limits. They include meditation practices that create space between stimuli and responses, allowing conscious choice rather than automatic reaction.

Psychedelic integration protocols that extract and integrate insights from expanded states of consciousness when used legally and responsibly. Flow state cultivation that accesses peak performance consciousness where limitations temporarily suspend.

Glitch state exploitation that uses naturally occurring altered states for rapid programming upgrades, and peak experience integration that captures and maintains insights from moments of expanded awareness.

I regularly use these tools to access consciousness beyond my normal waking state, where different types of insights and programming become possible. These expanded states often reveal limitations in my normal consciousness and provide access to upgraded code.

Tool 5: Teaching and Mentorship Systems

Master programmers develop the ability to help others upgrade their programming. This includes knowledge transfer systems that can effectively communicate programming concepts to others who are ready to learn.

Mentorship protocols that can guide others in their programming journey without doing the work for them. Example-setting algorithms that demonstrate programming principles through your own life and success, showing what's possible rather than just telling.

Support systems that provide ongoing guidance to others learning programming, offering feedback and encouragement, and community-building programming that creates environments where people can learn and grow together, amplifying individual development through collective support.

Teaching others to program their reality serves multiple purposes: it deepens your own understanding, you truly master what you teach. It fulfills the natural desire to contribute once you've achieved success. It creates community of conscious programmers supporting each other's growth, and it contributes to the evolution of human consciousness by spreading these capabilities.

I now spend significant time teaching these principles to others, not because I've completely mastered them myself (mastery is ongoing), but because teaching deepens my own practice while helping others access these tools.

The Master Programmer's Journey

Becoming a master programmer isn't an event, it's a developmental journey through predictable stages.

Stage 1: Recognition (Weeks 1-4)

You're realizing that you've been running programming rather than just experiencing random life events. You're beginning to see patterns in your thoughts, emotions, and behaviors. You're recognizing that your limitations might be code rather than fixed reality.

At this stage, you're developing basic awareness that programming exists and affects your life. You're not yet able to modify code effectively, but you can observe it running. This observation itself begins to create space between programming and conscious choice.

Stage 2: Responsibility (Months 2-3)

You're accepting that you're responsible for your code and its outputs. You're moving from "life happens to me" to "I generate my experience through my programming." You're taking ownership of your results rather than blaming external factors.

This stage can feel uncomfortable as you recognize that your current struggles are outputs of your current code. But this discomfort is the price of power, once you accept responsibility, you gain the authority to change things.

Stage 3: Learning (Months 4-6)

You're actively learning debugging and programming techniques. You're studying how consciousness works, how beliefs generate experiences, how patterns can be modified. You're experimenting with basic programming upgrades and observing results.

At this stage, you're building your programming skillset through study, practice, and experimentation. You're not yet masterful, but you're systematically developing capabilities.

Stage 4: Practice (Months 7-12)

You're consistently implementing programming upgrades across all life areas. You're getting results that prove your programming changes create real-world changes. You're developing confidence in your ability to modify your code and generate different outcomes.

This stage is where theoretical understanding becomes practical capability through consistent application. You're proving to yourself that this works through direct experience.

Stage 5: Mastery (Year 2-3)

You're operating as a conscious programmer most of the time. Programming upgrades have become natural rather than effortful. You're creating significant results through your upgraded code. You're helping others learn to program their reality.

At this stage, master programmer consciousness has become your default rather than something you have to work to access. You naturally see reality as programmable and yourself as the programmer.

Stage 6: Teaching (Year 3+)

You're contributing to others' development by sharing what you've learned. You're creating resources, mentoring others, and building community around conscious reality programming. You're using your mastery to elevate human consciousness.

This stage represents maturity where your focus expands beyond personal optimization to collective elevation. You've achieved sufficient mastery that your primary interest becomes helping others develop similar capabilities.

Advanced Master Programmer Practices

Once you've developed basic master programmer capabilities, you can implement sophisticated practices that continuously enhance your effectiveness.

Practice 1: Daily Programming Sessions

Establish a daily practice for consciously programming your reality. This isn't just morning meditation, it's active consciousness programming work.

My daily session includes reviewing current code execution, what programming ran yesterday and what results it generated. Debugging identified limitations, what beliefs created unwanted outcomes and how to upgrade them. Installing chosen programming, consciously rehearsing upgraded code through visualization and mental practice. Setting programming intentions, choosing what code to run today and committing to it. Reviewing long-term programming goals, ensuring daily programming aligns with overall development trajectory.

This fifteen to thirty minute daily practice ensures I'm programming myself consciously rather than letting random reinforcement shape my code. It's the highest-leverage time investment possible, everything else in my life flows from the code I'm running.

Practice 2: Weekly System Reviews

Each week, conduct a comprehensive review of your programming and its outputs. This includes assessing code execution quality, how consistently did you run chosen programming versus defaulting to old code. Analyzing results and patterns, what outcomes did your programming generate and what patterns do you notice. Identifying bugs and limitations, what programming created problems or limited results. Planning upgrades and optimizations, what code modifications would improve next week's results. Adjusting programming strategy, based on this week's data, what changes to your approach are needed.

This weekly review creates feedback loops where your results inform your programming decisions, allowing continuous optimization based on actual performance rather than theoretical assumptions.

Practice 3: Monthly Programming Upgrades

Each month, implement a significant programming upgrade in one major life area. This provides focused intensive work that creates substantial improvements without trying to upgrade everything simultaneously.

The monthly upgrade cycle includes selecting target area, which life domain most needs optimization this month. Comprehensive code audit, thoroughly documenting current programming in that area. Designing upgraded code, creating specific new beliefs, thoughts, and patterns. Installation protocol, systematic daily practice for implementing the upgrade. Testing and adjustment, monitoring results and refining the code based on feedback. Integration into overall system, ensuring the upgrade works harmoniously with other programming.

This focused monthly approach creates substantial improvements over time without overwhelming yourself trying to upgrade everything at once.

Practice 4: Quarterly Strategic Planning

Every three months, step back from daily programming to conduct strategic assessment and planning. This includes reviewing progress over the past quarter, what's improved, what hasn't, why. Assessing current programming architecture, does your overall system still serve your goals or need restructuring. Planning major upgrades, what significant changes are needed in the coming quarter. Aligning with long-term vision, ensuring your programming trajectory leads toward your ultimate goals. Optimizing support systems, what resources, relationships, or tools would enhance your programming work.

This quarterly strategic view prevents getting lost in daily details while ensuring you're progressing toward your larger vision rather than just optimizing within limited parameters.

Practice 5: Annual System Overhaul

Once per year, conduct a comprehensive assessment and major version upgrade of your entire reality operating system. This includes complete system audit, thoroughly reviewing all programming across all life areas. Results assessment, evaluating what outcomes your programming has generated over the year. Major upgrade planning, designing significant enhancements for the coming year. System architecture refinement, restructuring how your various programming systems interact. Vision clarification, updating or refining your long-term goals and values. Legacy planning, considering what you want to create and contribute long-term.

This annual practice ensures continuous evolution rather than plateauing at any particular level of development.

The Master Programmer's Responsibility

With master programmer capabilities comes significant responsibility. This power can be used for genuine growth and contribution, or it can be misused in ways that ultimately create suffering.

Responsibility 1: Ethical Programming

Master programmers must ensure their code creates positive outcomes not just for themselves but for others in their reality. Programming that succeeds at others' expense, that manipulates or harms others, or that creates value for yourself while destroying value for others ultimately creates problems even if it generates short-term results.

Ethical programming includes impact assessment, considering how your programming affects others and the world. Value alignment, ensuring your code aligns with your highest values and ethical standards. Harm prevention, avoiding programming that creates negative consequences for others. Contribution focus, using success to make positive contributions beyond yourself. Responsibility acceptance, taking accountability for the effects of your programming on others.

I regularly examine: "Does my programming create value for others or just extract it? Does my success contribute to collective wellbeing or just individual gain? Are there negative externalities from my code that I need to address?"

Responsibility 2: Continuous Improvement

Master programmers commit to ongoing development rather than treating programming as a one-time achievement. Your code needs continuous optimization to meet changing circumstances and growing capabilities.

This includes learning systems that continuously acquire new knowledge and skills, keeping your programming current and expanding. Adaptation protocols that adjust to changing circumstances, ensuring your code remains effective as situations evolve. Optimization algorithms that continuously improve performance, preventing complacency and stagnation. Growth mindset maintenance that sees challenges as development opportunities rather than threats. Evolution commitment that treats programming as lifelong practice rather than destination.

I view myself as perpetually in development, not because I'm flawed or inadequate, but because evolution is the natural state of conscious beings. Stagnation is death; growth is life.

Responsibility 3: Teaching and Mentorship

As you develop master programmer capabilities, you become responsible for helping others access these tools. Knowledge that could benefit humanity becomes responsibility to share when you possess it.

This includes knowledge transfer, sharing programming concepts and techniques with others ready to learn. Mentorship provision, guiding others in their programming journey without doing their work for them. Example setting, demonstrating through your life what's possible with upgraded programming. Support offering, providing ongoing guidance and encouragement to others developing these skills. Community building, creating environments where people can learn and practice together.

I now see teaching these principles as responsibility, not just option. The tools that transformed my life from poverty and chaos to success and fulfillment could help countless others. Keeping them to myself would be unethical hoarding of valuable resources.

Responsibility 4: System Maintenance

Master programmers maintain their systems carefully to prevent bugs, degradation, and security vulnerabilities. Neglected systems develop problems regardless of how well they were initially designed.

This includes regular auditing, consistently reviewing your programming effectiveness and health. Bug prevention, identifying and addressing potential problems before they become serious. Performance monitoring, tracking how well your system operates and maintaining high standards. Security maintenance, protecting your consciousness from harmful external programming and maintaining boundaries. Optimization practice, continuously improving your system's efficiency and effectiveness.

I treat system maintenance as seriously as any other commitment. My consciousness is the foundation of everything else in my life, neglecting its maintenance would undermine everything built upon it.

Responsibility 5: Innovation and Evolution

Master programmers explore new possibilities and expand human understanding of consciousness programming. You become responsible for pushing boundaries of what's possible and contributing to collective knowledge.

This includes experimental programming, trying new approaches and documenting results. Boundary exploration, testing limits of consciousness capabilities. Creative development, developing new programming techniques and approaches. Possibility expansion, discovering new capabilities and applications. Knowledge contribution, sharing discoveries that advance collective understanding.

I experiment with advanced techniques and share what works, contributing to the evolving understanding of consciousness programming. Individual innovation serves collective evolution.

The Ultimate Responsibility: Legacy Creation

The ultimate responsibility of master programmers is using their capabilities to create lasting positive impact. Once you've mastered your own programming and created the life you want, the question becomes: What will you do with these capabilities to contribute to something larger than yourself?

This is where personal development transforms into transpersonal contribution. Your programming mastery becomes a tool for addressing important problems, helping others develop their capabilities, creating systems that continue benefiting people long after you're gone, and advancing human consciousness and potential.

I didn't develop these capabilities just to have a comfortable life, though that's valuable. I developed them to contribute to solving meaningful problems, to help others who are trapped in limitation like I once was, to create resources and systems that will help people long after I'm gone, and to demonstrate what's possible when humans take conscious control of their reality programming.

Your legacy is what you create that matters beyond your own life. What will you build with your master programmer capabilities? What problems will you solve? Who will you help? What will you contribute that continues generating value after you're gone?

These aren't just philosophical questions. They're the ultimate test of master programmer consciousness. Have you transcended purely self-focused programming to create code that serves collective wellbeing? Are you using your mastery for contribution as well as personal benefit?

The journey from unconscious programming to master programmer consciousness is ultimately a journey from victimhood to authorship, from limitation to possibility, from reactive to creative, from self-focused to service-focused. Each stage serves its purpose, but the ultimate expression of mastery is using your capabilities to contribute to something larger than yourself.

You've learned to hack the simulation. Now what will you create?

Chapter 10: Your New Reality Awaits

As I write these final words, I'm sitting in my home office, looking out at a life that would have seemed utterly impossible to the version of me who slept in tents, scrounged for food, and believed that poverty and struggle were just the reality "people like me" experienced. The contrast still strikes me sometimes with such force that I have to pause and really absorb it: this life I'm living now, this reality I'm experiencing, I created it. Not through luck, not through special advantages, not through being chosen by external forces. I created it by learning to program my reality consciously rather than running inherited code unconsciously.

Everything I've shared in this book, from the basic understanding of life as code to the advanced techniques of master programming, is available to you right now. You don't need special credentials, perfect circumstances, or anyone's permission to start rewriting your reality. You just need to understand that the power has always been in your hands, you simply didn't know it was there or how to use it.

The simulation you're living in right now is running code that was installed by others, generated by circumstances, and accepted by you before you knew you had a choice. Now you know. Now you understand that every limitation, every struggle, every "unchangeable" aspect of your life is actually just code that can be debugged, upgraded, and optimized.

Your poverty isn't permanent, it's just the output of scarcity programming that can be rewritten into abundance code. Your relationship problems aren't character flaws, they're the result of outdated connection algorithms that can be updated. Your career limitations aren't fixed by your background, they're generated by professional programming that can be optimized. Your health struggles aren't genetic destiny, they're the output of vitality code that can be upgraded. Your emotional challenges aren't personality disorders, they're the result of feeling management systems that can be enhanced.

Every single aspect of your life is programmable. Every limitation is optional. Every problem is solvable through better code.

The Thirty-Day Reality Hacking Challenge

To help you begin this transformation immediately, I'm giving you a comprehensive thirty-day challenge that will start rewiring your simulation. This isn't theory or inspiration, it's practical daily action that will begin generating measurable changes in your reality within weeks.

Days 1-7: Recognition Phase

The first week focuses on developing awareness that you're running code rather than just experiencing fixed reality. This recognition is the foundation everything else builds on.

Daily Practice (15 minutes):

Each morning, spend ten minutes observing your thoughts as code execution. Notice thoughts arising without identifying with them completely. Recognize automatic thoughts, recurring patterns, and habitual responses as programming running rather than truth. Document at least three examples daily of code you caught running.

Throughout the day, maintain awareness: "Is this thought/feeling/behavior code executing, or is this a conscious choice?" This simple question begins creating space between programming and conscious awareness.

Each evening, journal about patterns you noticed: What code ran most frequently today? What results did that code generate? What patterns do you recognize across multiple days? What programming would you like to upgrade?

By day seven, you should have clear documentation of your active programming patterns, recognition of code running in real-time at least several times daily, and beginning awareness that limitations might be programming rather than reality.

Days 8-14: Debugging Phase

The second week focuses on identifying and beginning to debug your most limiting code. You're not completely rewriting everything yet, just starting to interrupt automatic patterns.

Daily Practice (20 minutes):

Each morning, choose one specific limiting belief to debug today. Trace it back to its origins: When did you first believe this? What circumstances installed this code? How has this belief served or protected you? Is it still serving you now, or is it now limiting you?

Throughout the day, catch that specific belief running and consciously challenge it: When the thought arises, notice it: "There's that code again." Question it: "Is this actually true, or is this just programming?" Replace it consciously with upgraded code: "I used to believe X, but now I'm learning Y."

Each evening, document: How many times did you catch the limiting belief today? How successfully did you challenge and replace it? What results came from running upgraded code versus old code? What resistance or difficulty did you experience?

Choose a different belief to debug each day, or work with the same belief all week if it's particularly persistent.

By day fourteen, you should have actively debugged at least seven different limiting beliefs, developed the skill of catching code as it runs and consciously interrupting it, experienced at least some instances of different results from running different code, and built confidence that programming is actually modifiable.

Days 15-21: Installation Phase

The third week focuses on actively installing upgraded programming, not just debugging old code. You're beginning to run new code consciously and consistently.

Daily Practice (25 minutes):

Each morning, choose upgraded programming to install today. Write out the new belief explicitly. Visualize yourself operating from this belief, make it vivid and emotionally real. Practice thinking thoughts that align with this belief. Feel what it would feel like if this belief were already true.

Throughout the day, take at least three actions aligned with your upgraded programming: Make one decision from the new belief instead of old code. Respond to one situation how your upgraded programming would respond. Notice one opportunity your new code reveals that old code would have filtered out.

Each evening, document: What actions did you take from upgraded programming today? What results did those actions generate? How did it feel to act from new code versus old code? What evidence did you collect that supports your upgraded belief? What resistance or sabotage did you notice?

By day twenty-one, you should have actively installed upgraded programming in multiple life areas, taken consistent actions from new code rather than just thinking about it, experienced concrete results that prove upgraded code generates different outcomes, and developed some comfort with new programming even though it's not yet automatic.

Days 22-28: Integration Phase

The fourth week focuses on integrating your upgrades so they start becoming automatic rather than requiring constant conscious effort.

Daily Practice (30 minutes):

Each morning, review all the upgraded programming you've been installing. Spend fifteen minutes visualizing your entire day operating from this upgraded code. See yourself making decisions, responding to situations, and interacting with others from your new programming. Make it as detailed and real as possible.

Throughout the day, maintain continuous awareness of what code you're running: Am I operating from upgraded programming or old code right now? If old code, consciously switch to upgraded code. If upgraded code, acknowledge and reinforce: "Yes, this is who I'm becoming."

Each evening, conduct comprehensive review: What percentage of the day did you operate from upgraded versus old code? What triggers caused reversion to old code? How quickly could you recognize and correct when old code ran? What's becoming more automatic versus still requiring effort?

Also this week, begin modifying your environment to support upgraded programming: Change physical reminders, remove what triggers old code, add what reinforces new code. Adjust social interactions, spend more time with people who reflect your upgraded programming. Modify information diet, consume content that reinforces new code, eliminate content that reinforces old code.

By day twenty-eight, you should notice upgraded programming feeling more natural and less forced, automatic operation from new code in at least some situations, reduced effort required to maintain upgraded programming, and environmental supports reinforcing new code rather than triggering old code.

Days 29-30: Optimization and Commitment Phase

The final two days focus on optimizing what you've learned and committing to ongoing practice.

Day 29 Practice:

Conduct comprehensive review of the entire month: What programming have you upgraded? What results have you created through upgraded code? What's still challenging or inconsistent? What patterns do you notice in successful versus struggling areas? What practices have been most effective? What commitments will you make going forward?

Document your transformation in detail, not just what changed, but how the changes happened and what practices created them. This documentation serves both as evidence of what's possible and as a manual for future upgrades.

Day 30 Practice:

Design your ongoing programming practice: What daily programming routine will you maintain? What weekly and monthly practices will you establish? What accountability or support systems will you create? What areas will you focus on next? What's your three-month programming roadmap?

Make explicit commitments to ongoing practice. Programming maintenance requires consistent work, not one-time fixes. Decide now what practices you'll maintain beyond this thirty-day foundation.

By day thirty, you should have completed a full cycle of recognition, debugging, installation, and integration; experienced measurable changes in at least some life areas through upgraded programming; developed a sustainable daily practice for ongoing programming work; and made clear commitments to continuing this work long-term.

The Long-Term Programming Strategy

The thirty-day challenge establishes your foundation, but real mastery develops over months and years of consistent practice. Here's your roadmap for continued development.

Months 2-3: Foundation Solidification

Continue and deepen your basic practices. The programming you installed during month one needs reinforcement to become truly stable. Many people make good progress in the first month, then revert when they stop practicing consistently.

Focus on making upgraded programming your actual default rather than something you have to consciously maintain: Practice upgraded code until it feels natural, not forced. Strengthen new neural pathways through consistent repetition. Eliminate or minimize environmental triggers for old code. Build support systems that reinforce new programming. Document ongoing results that prove new code works better.

By month three, upgraded programming should feel natural in at least some contexts, old code should activate less frequently and be easier to interrupt when it does, and you should have consistent results proving that programming changes create real-world changes.

Months 4-6: System Expansion

Systematically expand upgraded programming to all major life areas. You've probably focused on one or two areas most intensely, now consciously upgrade everything.

Each month, focus intensive work on one major area: Month four: upgrade the life area causing most limitation or suffering. Month five: upgrade the area with most potential impact if optimized. Month six: upgrade the area you've been avoiding or neglecting.

Use the same process for each area: identify current code, debug limitations, install upgrades, integrate through action, document results.

By month six, you should have upgraded programming running across all major life domains, integrated system where all programming works together coherently, and substantial improvements in multiple areas of life simultaneously.

Months 7-9: Optimization and Refinement

Fine-tune your programming for maximum effectiveness. You've installed upgrades, now optimize them for better performance.

Focus on making good code excellent: Increase consistency, run upgraded code even under stress or challenge. Improve quality, make upgraded code more sophisticated and nuanced. Enhance integration, ensure all programming systems work together seamlessly. Build resilience, make upgrades resistant to regression. Develop mastery, move from conscious effort to unconscious competence.

By month nine, upgraded programming should be automatic in most situations, only requiring conscious attention during unusual stress or challenge. Results should be consistently positive across all areas. You should feel genuine mastery over your programming rather than still struggling with it.

Months 10-12: Mastery Development

Develop master programmer capabilities and begin helping others learn these tools.

Focus on advanced practices: Real-time code modification during situations. System architecture design and optimization. Meta-programming, code that improves other code. Teaching others what you've learned. Contributing to collective understanding of consciousness programming.

By month twelve, you should operate as conscious programmer most of the time, seeing reality as programmable and yourself as the programmer naturally. You should have created substantial results through upgraded programming, measurably better outcomes across all life areas. You should be helping others understand and apply these tools, and you should have clear vision for ongoing development and contribution.

The Core Programming Principles

As you continue this work over months and years, remember these fundamental principles that make effective programming possible.

Principle 1: Responsibility

You are responsible for your programming and its results. No one else can fix your code for you. This isn't blame, it's power. Responsibility equals authority to change things.

Accept complete responsibility for your current results, recognizing they're outputs of your current code. Take ownership of debugging and upgrading your programming rather than waiting for external solutions. Acknowledge that your future results will be determined by the code you choose to run now.

Principle 2: Consistency

Programming changes require consistent practice and reinforcement to become permanent. Occasional inspiration or understanding creates temporary changes. Daily practice creates lasting transformation.

Practice upgraded code daily, not just when you feel motivated. Maintain your programming work even when you don't see immediate results, neural rewiring takes time. Establish sustainable routines rather than relying on motivation or willpower. Recognize that consistency matters more than intensity, fifteen minutes daily beats occasional marathon sessions.

Principle 3: Patience

Programming optimization is gradual process that takes time and persistence. You're rewiring neural pathways developed over years or decades. Real change happens through accumulated practice, not instant transformation.

Don't expect overnight miracles, celebrate incremental progress. Allow time for new programming to stabilize before judging its effectiveness. Persist through the uncomfortable phase when new code feels forced and unnatural. Trust that consistent practice creates results even when progress isn't immediately visible. Recognize that sustainable transformation takes months and years, not days and weeks.

Principle 4: Courage

Upgrading your programming often requires stepping outside comfort zones and facing fears. New code feels uncomfortable before it feels natural. Growth requires courage to act before you feel ready.

Take actions your new programming requires even when they feel scary. Face the discomfort of old identity dissolving and new identity forming. Risk failure and judgment by trying new approaches. Stand in uncertainty while new patterns stabilize. Trust yourself to handle challenges that arise from changing your code.

Principle 5: Integrity

Your programming should align with your deepest values and create positive outcomes for yourself and others. Success through programming that violates your values creates internal conflict. Impact matters, not just personal benefit.

Ensure your upgraded code serves your authentic values, not just ego or external validation. Consider how your programming affects others and the world, not just yourself. Use your success to contribute to collective wellbeing, not just personal comfort. Maintain ethical standards as you develop power through programming mastery. Remember that sustainable success requires integrity, shortcuts that compromise values eventually create problems.

Principle 6: Continuous Learning

Programming mastery requires ongoing learning, experimentation, and refinement. What works at one level of development may need upgrading as you grow. Stay open to new approaches and techniques.

Continue studying consciousness, neuroscience, psychology, and systems thinking. Experiment with new programming approaches and techniques. Learn from both successes and failures, both provide valuable data. Stay curious about what's possible rather than assuming you know everything. Recognize that mastery is ongoing process, not destination, there's always a next level.

Principle 7: Service

Your programming capabilities should ultimately serve something greater than personal benefit. Contributing to others' growth and collective wellbeing creates sustainable fulfillment. Individual mastery finds meaning through transpersonal contribution.

Use your success as platform for helping others, not just consuming more. Share what you learn so others can benefit from your discoveries. Create systems and resources that continue helping people beyond your direct involvement. Recognize your development as part of human consciousness evolution. Find ways to contribute to solving meaningful problems in the world.

What This Makes Possible

If you implement these teachings consistently over the next year, what becomes possible? Let me paint a picture based on my experience and hundreds of others who've applied these tools.

In Your Career: You'll likely experience significant advancement, promotions, new opportunities, increased income, or successful transition to work you actually care about. Not through grinding harder, but through upgraded programming that makes you more valuable, helps you recognize and pursue better opportunities, positions you strategically, and creates results others notice and reward.

Typical outcomes include 50-100% income increases within the first year, career advancement that would normally take 3-5 years happening in 1-2 years, transition to more meaningful and fulfilling work, recognition and opportunities that previously seemed out of reach, and confidence and capability in professional contexts that felt intimidating before.

In Your Finances: You'll likely experience substantial improvement in your financial situation through upgraded money consciousness, better earning through improved value creation and positioning, smarter financial decisions from abundance code replacing scarcity code, reduced financial stress as your relationship with money transforms, and beginning wealth building as you move from consumption to investment thinking.

Typical outcomes include elimination or dramatic reduction of debt within 12-18 months, emergency funds and savings that provide real security, investment and wealth-building practices that compound over time, abundance mindset that attracts financial opportunities, and freedom to make choices based on values rather than just financial necessity.

In Your Relationships: You'll likely experience significant improvement in relationship quality and satisfaction through upgraded relationship algorithms that attract healthier connections, better communication from emotional mastery and reduced reactivity, deeper intimacy from authentic expression rather than people-pleasing, healthier boundaries that maintain your integrity while preserving connection, and resolution of recurring patterns that previously sabotaged relationships.

Typical outcomes include existing relationships improving dramatically or releasing relationships that require you to stay limited, attracting partners and friends who support your growth rather than preferring you limited, reduced relationship drama and increased genuine connection, comfort with intimacy and independence rather than anxious or avoidant patterns, and relationships as source of energy and growth rather than drain and struggle.

In Your Health: You'll likely experience substantial improvements in physical wellbeing through upgraded health consciousness, significantly increased energy and vitality, resolution of stress-related symptoms and chronic issues, improved relationship with your body from partnership rather than adversarial consciousness, better sleep, eating, movement, and self-care from changed priorities and beliefs, and physical capability that expands as your code stops limiting it.

Typical outcomes include 40-60% increase in daily energy levels, resolution of chronic symptoms you'd accepted as normal, measurable improvements in health markers and biomarkers, genuine enjoyment of your body rather than constant management of problems, and physical performance and capability that surprises you with what's possible.

In Your Emotional Life: You'll likely experience dramatic improvements in emotional wellbeing through emotional mastery rather than reactivity or suppression, significantly reduced anxiety, depression, or emotional volatility, ability to process difficult emotions skillfully rather than being overwhelmed, emotional resilience that strengthens through challenges rather than breaking, and access to positive emotions like joy, peace, and contentment that were previously elusive.

Typical outcomes include emotional baseline shifting from anxious or dissatisfied to peaceful or content, reduced emotional drama and reactivity in all relationships, ability to maintain chosen emotional states even during challenges, emotional experiences as information and energy rather than overwhelming forces, and genuine emotional freedom rather than feeling controlled by feelings.

Most Importantly: You'll experience a fundamental shift in how you relate to your life. You'll move from feeling like life happens to you to recognizing that you're creating your experience through your programming. You'll shift from limitation thinking to possibility thinking. You'll transition from hoping things will change to knowing you can change things by changing your code.

This shift in consciousness, from victim to creator, from reactive to proactive, from unconscious to conscious, is more valuable than any specific result. It's the foundation that makes everything else possible.

The Promise and the Warning

I want to be completely clear about what I'm promising and what I'm not.

What I Am Promising:

If you implement these teachings consistently, your life will improve measurably across multiple dimensions. You will create better results than you're creating now. You will develop capabilities you currently don't have. You will experience yourself and your reality differently. The changes will be substantial, often dramatic, and sustainable if you maintain the practices.

This works. I know because it transformed my life from poverty, chaos, and limitation to success, stability, and expanding possibility. I know because I've taught these tools to hundreds of people who've used them to create their own transformations. I know because the principles are grounded in solid neuroscience, psychology, and systems theory, not wishful thinking.

What I Am Not Promising:

I'm not promising that you'll get everything you want immediately. Programming changes take time, weeks for small changes to feel natural, months for major transformations to stabilize. I'm not promising that you won't have to do anything, these tools require consistent practice and action, not passive hoping.

I'm not promising that external circumstances won't matter, of course they matter. But your programming determines how you perceive circumstances, what opportunities you notice, how you respond, and therefore what results you create. I'm not promising that upgraded programming eliminates all challenges, life still includes difficulties. But you'll handle challenges more effectively and suffer less while navigating them.

I'm not promising that reality programming alone solves everything, sometimes you need medical care, therapy, financial advice, or other professional support. These tools enhance your capacity to use such support effectively, but they don't replace professional help when it's needed.

The Warning:

This work requires commitment, courage, and honesty. You can't just read this book and hope for changes, you must implement the practices consistently. You can't do this halfheartedly and expect full results, your level of commitment determines your level of transformation.

You'll encounter resistance, internal resistance as old programming fights to maintain itself, and external resistance as others react to your changes. You'll experience discomfort, new programming feels forced and fake before it feels natural. You'll face moments of doubt, when old patterns resurface, you'll question whether this works.

If you're looking for easy, comfortable, or quick solutions, this isn't it. If you're willing to do uncomfortable work consistently for months to create lasting transformation, this is the most powerful approach I've encountered.

The Final Invitation

The simulation has been running your entire life, executing code installed by your family, culture, and experiences, code you never consciously chose. That code has generated your current reality, whatever results, limitations, and struggles you're experiencing are outputs of the programming that's been running.

Most people live their entire lives without realizing they can change their code. They experience limitations as fixed reality rather than recognizing them as programming. They try to change outputs while leaving the underlying code unchanged, wondering why the same patterns keep repeating.

But you're different. You've read this book, which means some part of you already knows the truth: your current limitations aren't fixed features of reality. They're code, and code can be hacked.

You now have everything you need to begin: understanding that reality is programmable and your experience is generated by your code, tools for identifying limiting programming in every major life area, techniques for debugging beliefs and upgrading them systematically, frameworks for optimizing all your systems so they work together coherently, and knowledge of the path from unconscious reactivity to conscious mastery.

The only question remaining is: will you use what you've learned?

Will you complete the thirty-day challenge and begin actively reprogramming your reality? Will you maintain daily practice beyond initial motivation when results aren't yet visible? Will you persist through discomfort when new code feels unnatural and old code feels safe? Will you take responsibility for your programming and its outputs rather than blaming circumstances? Will you commit to ongoing mastery rather than treating this as one-time fix?

These aren't rhetorical questions. Your answers will determine whether this book creates lasting transformation or just provides temporary inspiration that fades.

Your Real Life Begins Now

Everything you've experienced up until this moment has been generated by code you didn't consciously choose. Your struggles, limitations, and stuck patterns, all outputs of inherited programming running automatically.

But this moment is different. This moment, you understand that you have admin access to your own consciousness. This moment, you know that every limitation is optional. This moment, you recognize yourself as the programmer rather than just the program.

What will you do with this knowledge?

You can close this book, feel inspired for a few days, then return to running your old code and wondering why nothing really changes. Most people choose this, it's safer, more comfortable, less demanding.

Or you can treat this as the actual beginning of your real life. You can start the thirty-day challenge tomorrow morning. You can maintain daily programming practice for the next year. You can systematically debug and upgrade every limiting pattern you've been running. You can discover what you're actually capable of when your programming stops interfering.

The code is waiting to be rewritten. The simulation is responsive to your programming. Your real potential is there, underneath the limiting code that's been blocking it.

I can't make you do this work. I can't force you to practice consistently or persist through discomfort. I can't give you someone else's transformation, you have to create your own.

But I can tell you this with absolute certainty: if you implement these teachings with real commitment, your life one year from now will be so different you'll barely recognize it. Not because of magic, but because upgraded programming generates upgraded results. Change the code, change the reality. It really is that straightforward.

The simulation has been waiting your entire life for you to discover you're the programmer. That discovery happened when you opened this book. Now the real work begins, actually programming your reality consciously rather than running inherited code unconsciously.

Your transformation starts with a single commitment: "I am the programmer of my reality, and I choose to program consciously."

Make that commitment. Start the thirty-day challenge tomorrow. Maintain the practices. Debug your limitations. Install upgrades. Optimize your systems. Become the master programmer of your existence.

Your current reality is just one possible output from one possible program. Infinite other realities are available through different code. You get to choose what program to run.

What reality will you create?

The simulation is listening. Your code is ready to be written. Your real life is about to begin.

Welcome to conscious reality programming. Welcome to your actual potential. Welcome to the rest of your life as the master programmer of your own simulation.

The question was never whether you could hack your reality. The question was whether you would.

Now you know how. The only question remaining is: will you?

Your new reality awaits. Begin programming.

Acknowledgments

This transformation would not have been possible without the countless people who supported my journey from unconscious programming to conscious creation. To my patients who taught me that healing happens at the programming level. To my mentors who showed me what was possible when you understand consciousness as code. To my colleagues who supported my unconventional approach to medicine and life. To my family who accepted my transformation even when it challenged their own programming, and to everyone who will read this book and use these techniques to create their own reality upgrades.

The simulation responds to whatever programming you choose to run. Choose wisely. Code consciously. Create magnificently.

Your new reality awaits.

About the Author

Dr. Pauline Stoltzner is a practicing nurse practitioner, consciousness programmer, and master of reality creation. She holds a PhD in Education with a specialization in Educational Technology and Design and is double board-certified in both family and psychiatric practice, with additional certifications in substance abuse treatment. She has successfully transformed her own life from poverty and chaos to professional success and personal fulfillment using the consciousness programming techniques described in this book.

She currently runs a successful medical practice, teaches at the University level as well as teaching consciousness programming to select students. She continues to research the intersection of consciousness, programming, and reality creation. She is the author of the acclaimed "Nurses Guide" series and lives with the understanding that reality is code, choosing to execute programming that creates abundance, health, meaningful relationships, and positive impact in the world.

This book contains the distilled wisdom of a lifetime spent learning to program consciousness and create reality, offering readers the same transformative techniques that changed everything for her.